Lecture Notes in Civil Engineering

Volume 177

Lecture Notes in Civil Engineering (LNCE) publishes the latest developments in Civil Engineering—quickly, informally and in top quality. Though original research reported in proceedings and post-proceedings represents the core of LNCE, edited volumes of exceptionally high quality and interest may also be considered for publication. Volumes published in LNCE embrace all aspects and subfields of, as well as new challenges in, Civil Engineering. Topics in the series include:

- Construction and Structural Mechanics
- Building Materials
- Concrete, Steel and Timber Structures
- Geotechnical Engineering
- Earthquake Engineering
- Coastal Engineering
- Ocean and Offshore Engineering; Ships and Floating Structures
- Hydraulics, Hydrology and Water Resources Engineering
- Environmental Engineering and Sustainability
- Structural Health and Monitoring
- Surveying and Geographical Information Systems
- Indoor Environments
- Transportation and Traffic
- Risk Analysis
- Safety and Security

To submit a proposal or request further information, please contact the appropriate Springer Editor:

- Pierpaolo Riva at pierpaolo.riva@springer.com (Europe and Americas);
- Swati Meherishi at swati.meherishi@springer.com (Asia - except China, and Australia, New Zealand);
- Wayne Hu at wayne.hu@springer.com (China).

All books in the series now indexed by Scopus and EI Compendex database!

More information about this series at http://www.springer.com/series/15087

Chirla N. V. Satyanarayana Reddy · Shinji Sassa
Editors

Scour- and Erosion-Related Issues

Proceedings of ISSMGE TC 213 Workshop

Springer

Editors
Chirla N. V. Satyanarayana Reddy
Department of Civil Engineering
Andhra University College of Engineering
Visakhapatnam, India

Shinji Sassa
Soil Dynamics Group
PARI, National Institute of Maritime, Port and Aviation Technology
National Research and Development Agency
Yokosuka, Japan

ISSN 2366-2557 ISSN 2366-2565 (electronic)
Lecture Notes in Civil Engineering
ISBN 978-981-16-4782-6 ISBN 978-981-16-4783-3 (eBook)
https://doi.org/10.1007/978-981-16-4783-3

This Springer imprint is published by the registered company Springer Nature Singapore Pte Ltd.
The registered company address is: 152 Beach Road, #21-01/04 Gateway East, Singapore 189721, Singapore

Foreword

From the prehistoric ages, human being has been building the shelters for residential area as well as agriculture facilities nearby in either the coastal area or riverbank. The construction technology of countermeasure for scour and erosion has evolved from utilization of natural material such as soils, weed, and jute with incorporating wooden poles to artificial fibers called "geosynthetics" and steel wire and bamboo nets as well as hollow concrete blocks.

The proceeding volume entitled *Scour- and Erosion-Related Issues* is an outcome of ISSMGE TC 213 Workshop which was held on December 16, 2020, as a Preconference Workshop of Indian Geotechnical Society Annual Conference 2020 in Visakhapatnam, India.

The Indian Geotechnical Society is one of the leading and active societies among the ISSMGE Membership societies in terms of hosting three to four ISSMGE TC Workshops per year, outstanding journal publication, Indian Geotechnical Journal in Springer Publication, and international level of 3-day-long annual conference.

The reported technical case histories and theoretical, experimental, and numerical research results in this proceedings would be a useful reference material to solve the scour and erosion problems along the coastal, riverbank, and bridge foundation. I sincerely congratulate Prof. Shinji Sassa, Chairman of TC 213, and Prof. Chirla N. V. Satyanarayana Reddy, editors and the authors of the articles for their contribution in publishing this book.

Incheon, South Korea

Prof. Eun Chul Shin
Vice President of ISSMGE for Asia
Professor Emeritus, Incheon National University
Chairman of Korea Consultant, Ltd.

Preface

The book contains seven chapters contributed by highly reputed international experts from academia and field practice based on the keynote lectures delivered in the International Workshop on "Scour and Erosion" organized on December 16, 2020, by the International Society for Soil Mechanics and Geotechnical Engineering (ISSMGE) on behalf of the Technical Committee 213 and Indian Geotechnical Society in association with Andhra University, Visakhapatnam. The event was organized in a grand manner as the preconference workshop of the prestigious Indian Geotechnical Conference (IGC) 2020 and was attended by 368 participants from 18 countries. We thank the contributors of the chapters for sharing their rich research and professional experiences pertaining to various issues related to scour and erosion. The book provides useful information for understanding the relevant mechanisms and causes for scour and erosion of soils and to adopt appropriate measures for their mitigation.

The chapter contributed by Sassa reports the recent advances in the mechanics and countermeasures of scour and internal erosion. Also, it presents a concise review on the key issues of the tsunami/wave–seabed–structure interactions from geotechnical and hydrodynamic perspectives. The behavior of laterally loaded piles under scoured conditions at bridges is analyzed in Jie Han's chapter. An integrated analysis of the effect of scour on the performance of pile-supported bridges is presented considering hydraulic, geotechnical, and structural aspects.

Madhav's chapter addresses the issues related to coastal erosion and scour of rivers and streams through case studies. It reports mitigation of coast erosion of Khambhat with a combination of gabions and geosynthetics. Also, a case history of the site with non-homogenous soil profile is presented to illustrate the anomaly of the present practices on estimation of scour of bridge piers. The chapter on "Sustainable Hard and Soft Measures for Coastal Protection" by Sundar discusses the relative merits and demerits of various coast protection systems and their suitability in different conditions based on the experiences gained in the field of coastal protection measures along the coastline of India. Albert's article deals with the various issues related to the geotextile tube design and construction and presents two case studies of the application of geotextile tubes in coastal protection and land reclamation projects.

Paolo's chapter presents new design approach along with the performance limits of new type of mattress made of double-twisted wire mesh based on observation of the mattress stability with respect to flow through mattress layer, combining the stability of the stone and its ability to control soil erosion underneath. Also, it presents five case studies covering different solutions to demonstrate the performance and durability of the mattresses. Satyanarayana Reddy's chapter presents issues related to beach erosion and soil erosion effects on foundation stability in sloping ground and deep excavation failures based on case studies pertaining to Visakhapatnam city. The various remedial measures to check the erosion are discussed.

We hope that the readers of the book may deepen their understanding on the causes and effects of beach/sea coast erosion, river erosion, and subsoil erosion caused by surface and subsurface flow of water and the mitigation methods with the interesting case studies presented in the chapters.

It is hoped that the book will be well received by academicians, researchers, and practicing engineers worldwide to address the issues related to scour and erosion.

Finally, we thank the TC 213 Committee of ISSMGE and National Executive Committee of IGS of the term 2019–2020 for the support in bringing out the book.

Visakhapatnam, India
Yokosuka, Japan

Chirla N. V. Satyanarayana Reddy
Shinji Sassa

About the Editors

Dr. Chirla N. V. Satyanarayana Reddy is a Professor of Civil Engineering at College of Engineering, Andhra University, India and has 28 years of teaching, research and consultancy experience. His expertise and research interests are in the areas of reinforced soils, ground improvement, landfills, soil retention in excavations, deep foundations and forensic geotechnical engineering. He obtained B.Tech. in Civil Engineering from Nagarjuna University, M.Tech. in Geotechnical Engineering from IIT Madras, M.E (Structures) from Andhra University and Ph.D. from NIT Warangal. He has guided 8 Ph.D. scholars and 92 M.Tech. dissertations. He has more than 110 publications in various national and international journals and seminars/conference proceedings. He has organized more than 30 seminars, workshops and training programs.

He handled several research projects funded by DST, UGC and AICTE. He received the Engineer of the Year 2006 Award from Government of Andhra Pradesh and The Institution of Engineers (India) A.P. State Centre, Best Academician Award for the year 2014 from Andhra University and Andhra Pradesh Scientist Award 2020 in Civil Engineering Discipline from A.P. State Council of Science and Technology, Government of A.P. He has organized more than 30 seminars, workshops, and training programs. He is fellow of Indian Geotechnical Society, The Institution of Engineers (India) and Life member of Indian Roads Congress, Indian Concrete Institute, Indian Society for Technical Education, Indian Society for Rock Mechanics and tunneling Technology. He served as National Executive Committee member of Indian Geotechnical Society during the period 2013–2020. He served as a member in H-4 Committee on Embankment, Ground Improvement and Drainage Committee of Indian Roads Congress (IRC) for the term 2018–2020. He is serving as member of TC 213 on "Scour and Erosion" of International Society for Soil Mechanics and Geotechnical Engineering (ISSMGE).

Dr. Shinji Sassa is Head of Soil Dynamics Group and Research Director of International Research Center for Coastal Disasters, Port and Airport Research Institute, National Institute of Maritime, Port and Aviation Technology, National Research and Development Agency, Japan. He obtained his Dr. Eng. from Kyoto University. He is best known for his seminal works on wave-induced seabed liquefaction that have been extensively cited worldwide. His main research areas are Soil Dynamics, Geodynamics, Coastal and Offshore Geotechnics, Subaqueous Sediment Gravity Flows, and Ecological Geotechnics. He was an invited panelist, twice, at the 15th (2001) and 17th (2009) International Conference on Soil Mechanics and Geotechnical Engineering, ISSMGE, and served as a panelist leader at the UNESCO conference on Submarine Mass Movements and Their Consequences, 2011. He is the recipient of numerous distinguished awards, including the Commendations by the Prime Minister for Disaster Prevention Merit and by the Minister of Land, Infrastructure, Transport and Tourism as a representative of Technical Emergency Control Force, the Commendation for Science and Technology by the Minister of Education, Culture, Sports, Science and Technology, Outstanding Research Accomplishment Award, Best Paper Award thrice of Soils and Foundations, and Best Technical Development Award twice from Japanese Geotechnical Society, Presidential Award from PARI, and Special Prize of Infrastructure Maintenance Award for Technological Development of Suppressing Washout, Internal Erosion and Cavity Collapse. He was also a recipient of several outstanding review awards for *Coastal Engineering*, *Journal of Rock Mechanics and Geotechnical Engineering*, *Applied Ocean Research*, and *Journal of Waterway, Port, Coastal, and Ocean Engineering*. His Liquefaction Prediction and Assessment Paper was Top-Read Paper in ASCE Most Read Articles 2017. He is currently Chair of the ISSMGE Technical Committee on Scour and Erosion, Editor for *Landslides: Submarine Landslides and Tsunami*, and Vice Chairman of Soils and Foundations.

Mechanics and Countermeasures of Scour and Internal Erosion: Tsunami/Wave-Seabed-Structure Interaction

Shinji Sassa

Abstract The paper reports some recent advances in the mechanics and countermeasures of scour and internal erosion. It presents a concise review of the key issues on the tsunami/wave-seabed-structure interactions from geotechnical and hydrodynamic perspectives. I highlight here seepage erosion, effect of overflow and seepage coupling on scour of breakwaters, wave-induced liquefaction and scour protection around a monopile, internal erosion, cavity formation and collapse behind seawalls and quaywalls.

Keywords Scour · Internal erosion · Wave

1 Introduction

Tsunami- and wave-seabed-structure interactions have received increasing attention in recent years, following the devastating impact of 2011 off the Pacific coast of Tohoku earthquake tsunami, ongoing development of oceanic infrastructures such as offshore wind turbine foundations and the need for coastal disaster risk management. The mechanics and countermeasures of scour and internal erosion plays a pivotal role in such tsunami/wave-seabed-structure interactions. The theme is multidisciplinary in nature and hence requires geotechnical and hydrodynamic perspectives.

Here, I report some key issues in the mechanics and countermeasures of scour and internal erosion through tsunami/wave-seabed-structure interactions. The paper first presents tsunami-seabed-structure interactions by focusing on seepage erosion and the effect of overflow and seepage coupling on the scour of caisson breakwaters with its countermeasure. Wave-seabed-structure interaction is then presented by highlighting wave-induced liquefaction and scour protection around a monopile and the mechanics and countermeasure of internal erosion, cavity formation and collapse behind seawalls and quaywalls.

S. Sassa (✉)
Port and Airport Research Institute, National Institute of Maritime, Port and Aviation Technology, 3-1-1 Nagase, Yokosuka 239-0826, Japan
e-mail: sassa@p.mpat.go.jp

C. N. V. S. Reddy and S. Sassa (eds.), *Scour and Erosion Related Issues*, Lecture Notes in Civil Engineering 177,
https://doi.org/10.1007/978-981-16-4783-3_1

2 Tsunami-Seabed-Structure Interaction

Tsunami-seabed-structure interactions are schematically shown in Fig. 1, which shows tsunami-induced forces on caisson breakwaters and failure modes of the foundation. It is essential to reproduce a prototype-scale stress field in clarifying the instability of breakwater foundations under tsunami. Geo-centrifuge makes it possible and has proven effective in studying fluid-soil interaction problems [1, 2]. In fact, the role and importance of centrifuge testing in fluid-soil-structure interactions have recently been emphasized [3–6]. Here, I report the novel use of geo-centrifuge for studying tsunami-seabed-structure interactions involving the role of tsunami-induced overflow and seepage on scour and erosion. The seepage flow in the rubble mound becomes turbulent due to the high permeability of the mound as crushed rocks. The similitudes for such turbulent seepage flow as well as overflow were satisfied together with the mechanical similarity between the model and the prototype [5, 7].

2.1 Seepage Erosion

Tsunami-induced seepage can induce erosion in sandy ground underneath the rubble mound of caisson breakwaters. Experiments were performed under conditions where

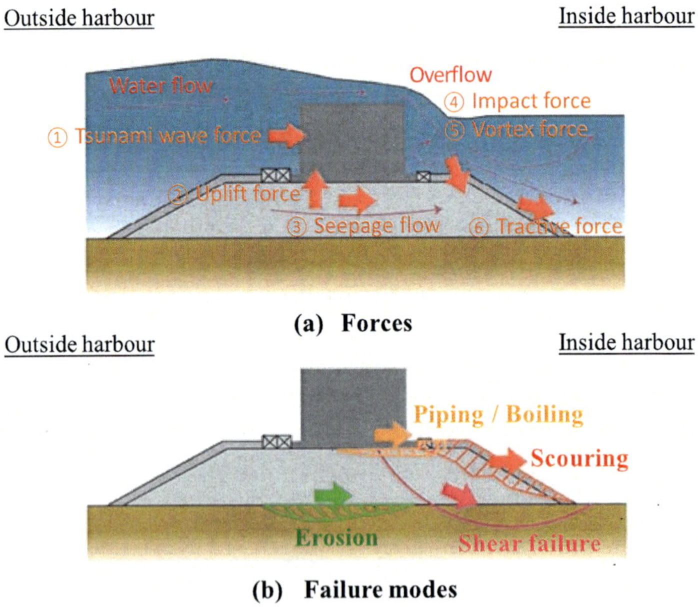

Fig. 1 Tsunami-induced forces on caisson breakwaters and failure modes of the foundation. Descriptions are added to Fig. 1 of [7]

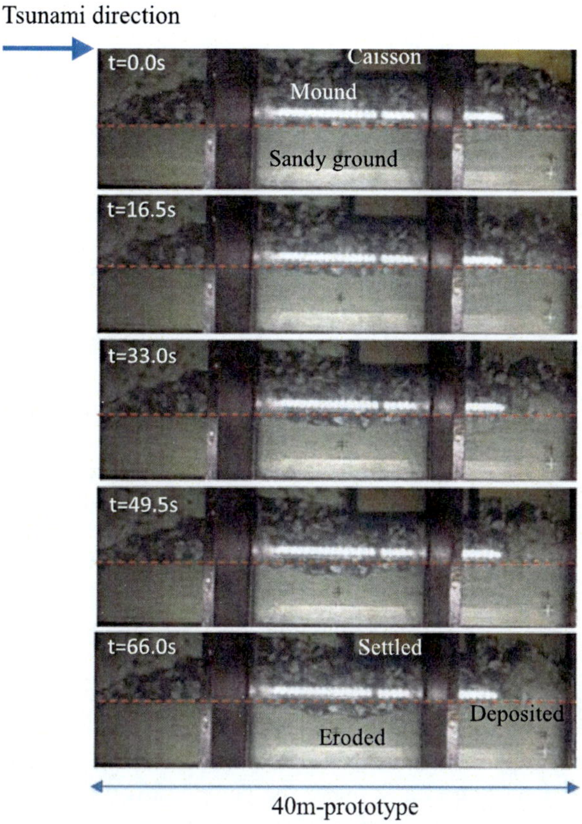

Fig. 2 Progress of seepage erosion. Descriptions are added to Fig. 7 of [7]. Settlement of the caisson was observed to occur with seepage erosion

no sliding and/or overturning of the caisson was allowed to occur in order to focus on the erosion process of the sandy ground. The experimental results on an actual breakwater (Omaezaki model) show that such seepage erosion progressed with time at the vicinity of the rubble/sand interface (Fig. 2). The eroded mass was deposited in the onshore side of the mound in accordance with the onshore direction of the seepage flow. Accordingly, the caisson settled as a consequence of the tsunami-induced seepage erosion.

2.2 *Effect of Overflow and Seepage Coupling on Scour*

The effect of overflow and seepage coupling on the scour of caisson breakwaters are described below. With reference to Fig. 1, the coupled overflow and seepage actions promoted the development of the mound scour significantly, causing bearing capacity

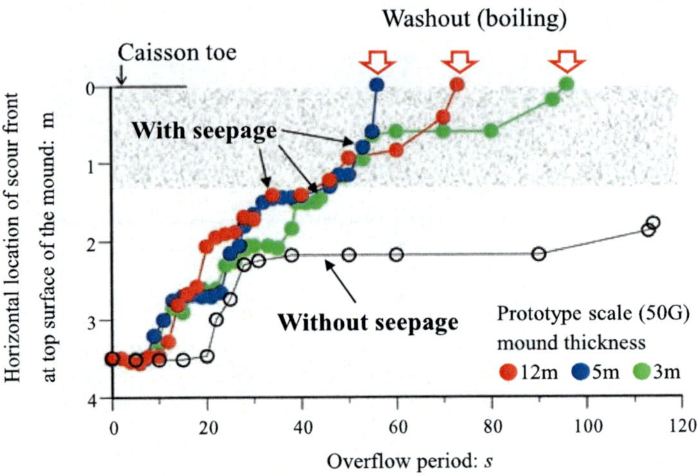

Fig. 3 Tsunami overflow and seepage coupling induced mound scour [5]. The shaded zone indicates the region where the caisson became unstable

failure of the mound, resulting in the total failure of the breakwater, which otherwise remained stable without the coupling effect, as shown in Fig. 3. The velocity vectors obtained from the high-resolution image analysis illustrated the series of such concurrent processes of the instability involving the scour of the mound/sandy seabed, bearing capacity failure and flow of the foundation, leading to the instability of the caisson breakwater under the coupled overflow and seepage (Fig. 4). The influence of placing an embankment (elevated mound) as a countermeasure was examined by employing a difference bank thickness. The embankment significantly suppressed the effect of such coupled overflow and seepage on scour, in accordance with a decreasing hydraulic gradient that manifested underneath the caissons, thus preventing bearing capacity failure, resulting in stabilization of the breakwater (Fig. 5).

Without a countermeasure, the scour front could reach the caisson toe, yielding the phenomenon of significant washout (boiling), as shown in Fig. 3, giving rise to the formation of a cavity underneath a remaining caisson. This was found to be consistent with what was observed following 2011 off the Pacific coast of Tohoku earthquake tsunami [5].

2.3 *Countermeasure*

Tsunami-resistant design of breakwaters has been developed (Fig. 6a, [8]). The embankment reinforces the caisson against the tsunami wave force and suppresses the effect of the coupled overflow and seepage on scour, as described above.

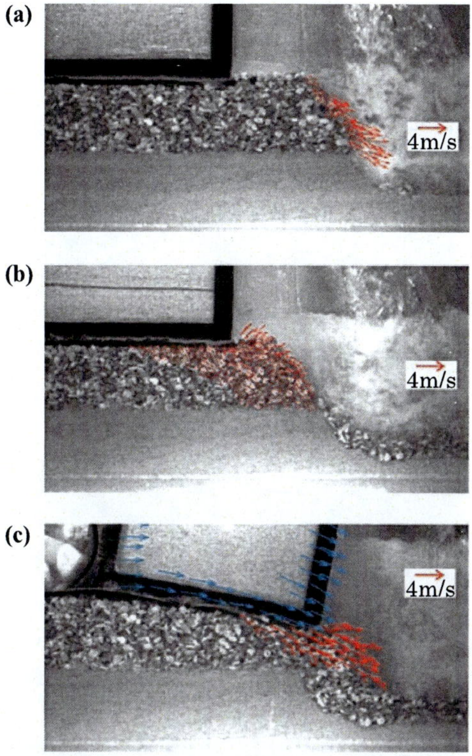

Fig. 4 Tsunami overflow and seepage coupling induced, **a** scour of the mound/seabed ground, **b** bearing capacity failure and **c** flow of the mound in the concurrent processes of the caisson instability, shown with velocity vectors [5]

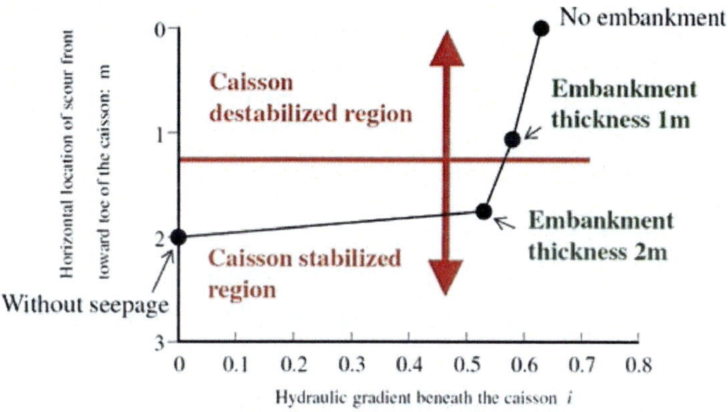

Fig. 5 Effect of the embankment and the influence of hydraulic gradient on the stability of the caisson due to coupled tsunami overflow and seepage [5]. The cases with no embankment and without seepage are shown for the purpose of comparison

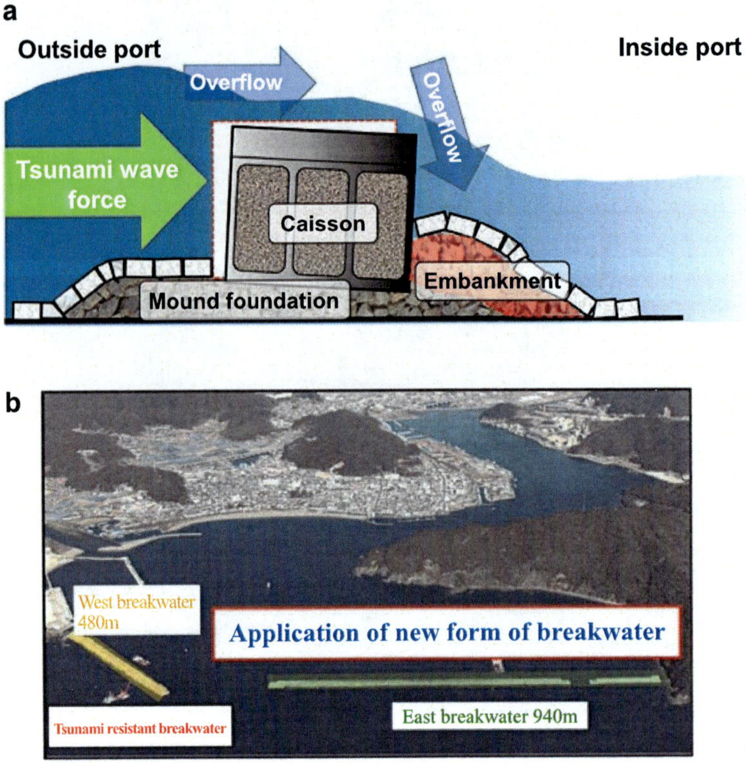

Fig. 6 **a** Tsunami-resistant breakwater with embankment: the sketch was drawn based on [8]. **b** Application of the tsunami-resistant breakwater with embankment to Suzaki Port, Japan

Such a new form of the breakwater with embankment has been brought into practice (Fig. 6b). It prepares for the megathrust earthquake tsunamis that are expected to occur in the near future.

3 Wave-Seabed-Structure Interaction

3.1 Wave-Induced Liquefaction and Scour Protection

Wave-induced seabed instability considerably affects the stability of coastal and offshore structures resting on/in the seabed. Representative forms and processes of the instability involve scour and seabed liquefaction. Here, I show that scour and wave-induced liquefaction are physically interconnected, affecting the stability of structures.

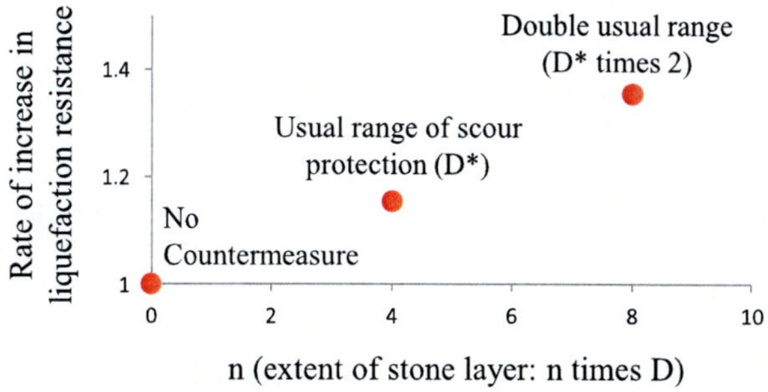

Fig. 7 Rate of increase in the seabed resistance to wave-induced liquefaction with scour protection around a monopile [6]

Wave-seabed-monopile interaction was studied in [6] by centrifuge wave testing, where the relevant scaling laws for wave-induced seabed liquefaction, i.e. time-scaling laws for fluid wave propagation and soil consolidation, as well as for mound scour were satisfied. Here, in a usual scour protection around a pier or monopile, the extent of the scour protection is 3–4 times the diameter D of the monopile [3]. The results shown in Fig. 7 demonstrate that scour protection increased the liquefaction resistance depending on the diameter ratios of the scour protection and the pile, however, it could not prevent the collapse of the monopile once liquefaction occurred. These results indicate the importance of soil stratification on the wave-seabed-structure interactions in light of wave-induced liquefaction and scour protection. Accordingly, wave-induced seabed liquefaction may need to be considered in a rational design of monopile foundations for offshore structures such as offshore wind turbine foundations.

3.2 *Internal Erosion*

This section highlights the processes of internal erosion, cavity formation and collapse behind seawalls and quaywalls. Typical cavity collapses behind seawalls are shown in Fig. 8. The photograph on the left shows the site of a fatal accident where a cavity formed in the backfill behind a caisson joint suddenly collapsed without giving any alarming signs to the soil surface. Investigations into the cause of the accident [9, 10] confirmed some key issues involved, as shown in Table 1. Namely, defects in sand covers allowed sand particles to be washed out through the caisson joint. The progress of internal erosion resulted in the occurrence and evolution of cavities under continued hydrodynamic forcing. Eventually, the cavity collapsed. The stability of such cavities in unsaturated granular backfills is of an essentially unsteady nature owing to ocean waves, tides, groundwater table fluctuations and precipitations. The key factors and processes involved are: propagation of fluctuating water pressure, wash out of sand particles, arch effect and the role of suction. The role of suction, i.e. negative pore water pressure relative to atmospheric pressure [11, 12], is of particular

Fig. 8 Cavity collapses behind seawalls

Table 1 Key issues in internal erosion, cavity formation and collapse behind seawalls/quaywalls

1. Defects in sand covers
2. Occurrence of cavities
3. Evolution of cavities
4. Collapse

⟺ The stability of cavities in unsaturated granular backfills is of an essentially unsteady nature

Ocean waves, Tides, Groundwater table fluctuations

Key factors and processes involved

◎ Propagation of fluctuating water pressure

◎ Wash out of sand particles

◎ Arch effect

◎ Role of **Suction**

⟺ No cavities formed in dry or saturated states of sands!

Flows of pore fluids (air and water)

importance, since no significant cavities would be formed in dry or saturated states of sands. This means that suction may control the lifetime of cavities in unsaturated granular backfills above groundwater levels accompanying flows of pore fluids (air and water) under complex hydro-environmental loading described above.

3.3 Countermeasure

Bi-layered filter (Fig. 9) is effective in preventing internal erosion, cavity formation and collapse behind seawalls and quaywalls. This is a dual protection concept such

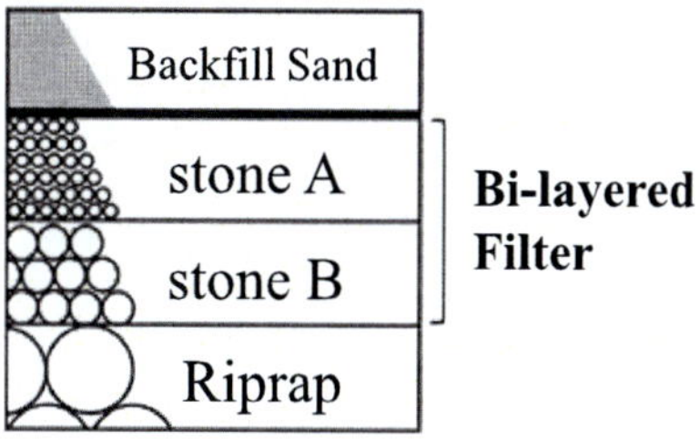

Fig. 9 Bi-layered filter for preventing internal erosion, cavity formation and collapse

that the washout, cavity and collapse are prevented from occurring even if sand covers/sheets fail. International standards on granular filters have generally been developed for use in dams and seabed [13–18]. However, backfill materials need three key issues to be considered: (1) Clogging/trapping function; (2) Stability against rubble stones; (3) Performance under various dynamic forcing such as waves, tides and earthquakes. Recent research has shown that grading of filters is important, particularly uniformity coefficient, U_{cF}, is a key for satisfying three functions stated above [19]. Consistent with technical standards of filters for dams and seabed, the following criterion can be of use for backfill materials to prevent internal erosion, cavity formation and collapse.

$$U_{cF} \geq 3,\ D_{F15}/D_{S85} \leq 5 \text{ or } D_{F50}/D_{S50} \leq 20 \tag{1}$$

With reference to Fig. 9, the above criterion applies to the upper filter layer beneath backfill sands. The application of such bi-layered filters to Naha Airport is shown in Fig. 10. The mitigation measure was taken to prevent internal erosion, cavity formation and collapse around box culverts subject to wave propagation and has proven necessary for a safe and maintenance-free operation of the important waterfront infrastructure.

4 Summary

The paper has presented some recent advances in the mechanics and countermeasures of scour and internal erosion through the tsunami/wave-seabed-structure interactions. Progress of seepage erosion beneath rubble mounds leads to the settlement of caisson breakwaters. Tsunami overflow and seepage coupling promote the development of scour of the mound/seabed ground, resulting in the total failure of caisson breakwaters. Tsunami-resistant design of breakwaters has been developed and brought into practice so as to prepare for future megathrust earthquake tsunamis. Wave-induced liquefaction and scour protection around a monopile are physically interconnected, affecting the performance of offshore wind turbine foundations. I highlighted some key issues in the mechanics and countermeasures of internal erosion, cavity formation and collapse behind seawalls/quaywalls. These may facilitate a better understanding of the physics involved pertaining to geohazard mitigation.

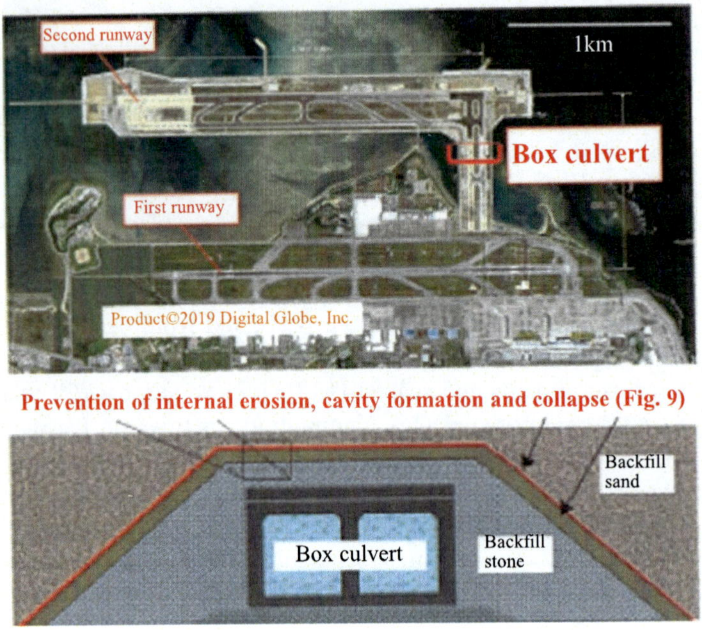

Fig. 10 Application of bi-layered filter to Naha Airport to prevent wave-induced internal erosion, cavity formation and collapse [20]

References

1. Sassa S, Sekiguchi H (1999) Wave-induced liquefaction of beds of sand in a centrifuge. Géotechnique 49(5):621–638
2. Sassa S, Sekiguchi H (2001) Analysis of wave-induced liquefaction of sand beds. Géotechnique 51(2):115–126
3. Sumer BM, Fredsøe J (2002) The mechanics of scour in the marine environment. Adv Ser Ocean Eng 17
4. Sumer BM (2014) Liquefaction around marine structures. Adv Ser Ocean Eng 39 (2014)
5. Sassa S, Takahashi H, Morikawa Y, Takano D (2016) Effect of overflow and seepage coupling on tsunami-induced instability of caisson breakwaters. Coast Eng 117:157–165
6. Miyamoto J, Sassa S, Tsurugasaki K, Sumida H (2021) Wave-induced liquefaction and instability of offshore monopile in a drum centrifuge. Soils Found 61(1):35–49
7. Takahashi H, Sassa S, Morikawa Y, Takano D, Maruyama K (2014) Stability of caisson-type breakwater foundation under tsunami-induced seepage. Soils Found 54(4):789–805
8. MLIT: Ministry of Land, Infrastructure, Transport and Tourism, Japan (2015) Tsunami-resistant design guideline for breakwaters
9. JSCE Coastal Engineering Committee (2002) Technical report on the 30 December 2001 accident in the Ohkura Artificial Sand Beach. The Japan Society of Civil Engineers (in Japanese)
10. Sassa S, Li F, Sekiguchi H (2005) Response of saturated/unsaturated gravelly sand to tidal fluctuations. ASCE Geotech Spec Publ 143:174–186
11. Bear J (1972) Dynamics of fluids in porous media. Dover, New York
12. Fredlund DG, Rahardjo H (1993) Soil mechanics for unsaturated soils. Wiley, New York
13. Fannin J (2008) Karl Terzaghi: from theory to practice in geotechnical filter design. J Geotech Geoenviron Eng 134(3):267–276
14. U.S. Army Corps of Engineers (1995) Design of coastal revetments, seawall, and bulkheads

15. U.S. Bureau of Reclamation (1998) Design of small dams
16. Indraratna B, Raut K, Khabbaz H (2007) Constriction-based retention criterion for granular filter design. J Geotech Geoenviron Eng 133(3):266–276
17. Recommendations of the Committee for Waterfront Structures Harbours and Waterways EAU (2012)
18. van Gent MRA, Wolters G (2015) Granular slopes with open filters under wave loading. Coast Eng 104:135–150
19. Kudai K, Sassa S, Yang S, Takada K, Suzuki K (2020) Processes of internal erosion, cavity formation and collapse under various soil and hydraulic conditions with their countermeasure using geotechnical filter. Rep Port Airport Res Inst 59(3):1–20
20. Mori H, Kinjo S, Ishigaki S (2019) Countermeasure against washout and collapse for the second runway of Naha Airport. Found Mag 47(12):76–78

Behavior of Laterally-Loaded Piles Under Scoured Conditions at Bridges

Jie Han

Abstract Scour often occurs around bridge piers and abutments, which are commonly supported by piles. This paper evaluates the effects of global scour and local scour on the behavior of laterally-loaded single piles and pile groups in sands and clays based on recent studies. The recent studies have been focused on stress history changes of remaining soils after scour, local scour-hole geometry and dimensions, and their effects on the behavior of laterally-loaded piles under scoured conditions. Experimental, numerical, and analytical studies have been conducted for these investigations and are discussed in this paper. Scour depth is identified as the most important influence factor, followed by scour width and slope. This paper also presents an integrated analysis of the effect of scour on the performance of pile-supported bridges that considered hydraulic, geotechnical, and structural aspects.

Keywords Deformation · Lateral load · Pile · Scour

1 Introduction

Scour is a process of erosion or removal of streambed or bank materials due to flowing water. Briaud [3] pointed out that erodibility depends on soil type or properties and water velocity. Scour can happen under different environments, e.g., around bridges and in marine environments. This paper will focus on scour around bridge piers and foundations. There are three common types of scour around bridge foundations: (1) general scour, (2) contraction scour, and (3) local scour. Arneson et al. [2] referred to the general scour as long-term degradation of a river bed. The general scour occurs even without any obstructions in a river channel and typically happens between large open spaces outside and between bridge foundations thus having shallow depths. The contraction scour happens when a water channel becomes narrow due to some

J. Han (✉)
The University of Kansas, Lawrence, KS 66045, USA
e-mail: jiehan@ku.edu

C. N. V. S. Reddy and S. Sassa (eds.), *Scour and Erosion Related Issues*,
Lecture Notes in Civil Engineering 177,
https://doi.org/10.1007/978-981-16-4783-3_2

obstructions (e.g., bridge abutments and piers), thus changing water flow velocity and direction, which increases the ability of water to remove soils (i.e., higher erodibility). The contraction scour typically occurs between bridge piers or between bridge piers and abutments. The general scour and the contraction scour often result in relatively uniform lowering of streambeds and they are together referred to as global scour in this paper. Around bridge piers, local scour can develop and extend below bridge foundations or pile caps if a pile foundation is used. Melville [24] provided the physics or mechanisms of local scour at bridge piers and illustrated local scour for six types of bridge piers: (1) uniform pier, (2) slab footing, (3) upwards tapering, (4) downwards tapering, (5) caisson foundation, and (6) pile foundation. Pile foundations and caisson foundations are two commonly used foundation types, especially for large-span bridges. In addition, pile foundations and monopiles have been increasingly used to support offshore structures and wind turbines, which may be subjected to scour. Piles under bridges and offshore structures are required to resist not only vertical loads but also lateral loads due to water flow, debris, and wind. Figure 1 shows the common types of scour around single piles and pile groups for analysis. Lin et al. [18] conducted a case history analysis of bridge failures due to scour and found that 63% of failures were caused by local scour. For pile groups, there are two types of local scour, which are referred to as group local scour and pile local scour in this

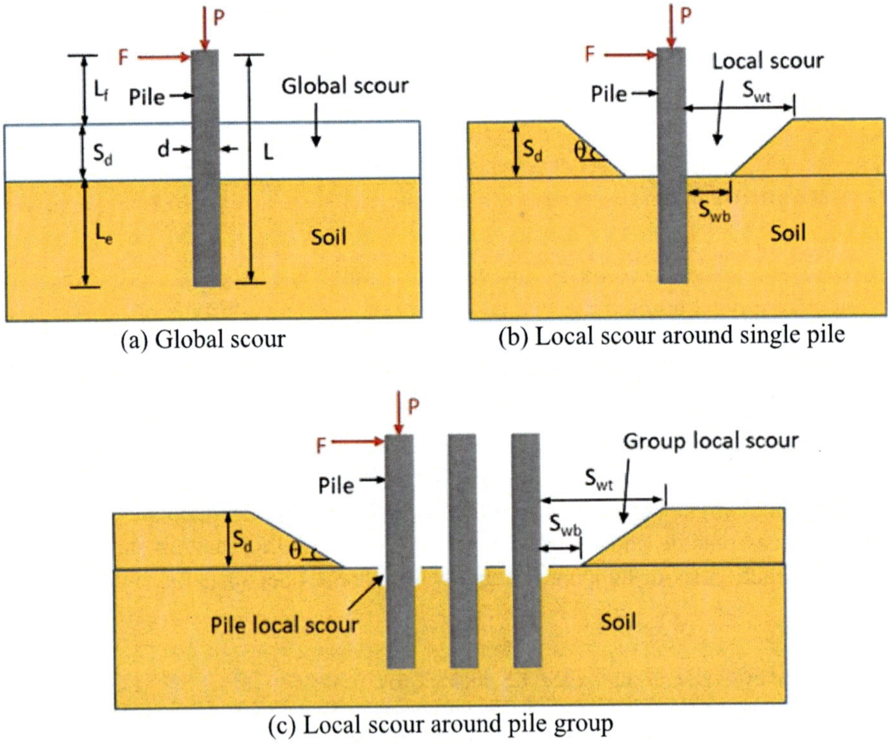

Fig. 1 Common types of scour around piles for analysis

paper. Global scour is commonly considered in the design of laterally-loaded single piles or pile groups under scoured conditions due to its simplicity and conservative nature in practice. Bridges on scoured foundations may also be subjected to seismic loading, which can have combined effects on the performance of the bridges [31]. Due to the page limit, the effect of seismic loading on the behavior of laterally-loaded piles under scoured conditions will not be discussed in this paper.

Wang et al. [31] conducted a comprehensive review of bridge scour in terms of mechanism, estimation (evaluation), monitoring, and countermeasures. Evaluation of scour at bridges involves hydraulic, geotechnical, and structural aspects. Depth of scour depends on many factors. Melville [24] grouped 13 specific influence factors into four key influence factors (flow rate, bed sediment, bridge geometry, and time). In the literature, a number of methods are available to estimate maximum scour depths around bridge piers and most of these methods are empirical and based on laboratory-scale data. Among these methods, the HEC-18 equation [2] is the most widely used, which considers local scour as a function of characteristics of riverbed material, bed configuration, flow characteristics, fluid properties, and the geometry of the pier and footing. Shape of scour hole depends on several factors. Arneson et al. [2] pointed out that the top width of a scour hole could range from 1.0 to 2.8 times the scour depth and depended on the bottom width of the scour hole and composition of the bed material but suggested that in practical applications the top width of the scour hole should be selected as twice the depth of the scour hole. Arneson et al. [2] also pointed out that the angle of repose of a cohesionless material in water is less than that in air. Around bridge foundations, Butch [6] found that the scour hole had an irregular shape, i.e., a steep slope in the upstream side and a gentle slope in the downstream side. For evaluating the effect of local scour on bridge piers and foundations, most researchers assumed a scour hole as an inversed truncated cone shape (e.g., Mostafa [25]; Li et al. [10]; Ismael [8]; Lin et al. [12, 13]; Zhang et al. [32]).

This paper presents recent studies on the behavior of laterally-loaded piles under scoured conditions and reviews the analyses done based on global scour and local scour with cone-shaped scour holes. In these analyses, the scour depth was assumed to be known. The behavior of laterally-loaded piles without any scour can be evaluated by different methods: (1) experimental, (2) analytical, and numerical. The behavior of laterally-loaded piles under scoured conditions can be evaluated by these methods as well. For example, Ismael [8] and Ismael and Han [9] conducted physical model tests to evaluate the effect of global and local scour on the lateral load-displacement curves of single piles. Lin et al. [12–15, 17], Zhang et al. [32], Lin and Wu [22], and Lin and Lin [20] developed analytical solutions to examine the effects of stress history of remaining soils and the scour-hole geometry and dimensions on the behavior of laterally-loaded single piles. The analytical solutions have been mostly based on the *p-y* curve concept (*p* is the lateral soil reaction and *y* is lateral pile displacement) because it is easier to understand and adopt. The *p-y* curve is to describe the interaction between pile and soil under loading by a series of linear or non-linear springs. Mostafa [25] and Li et al. [10] adopted numerical methods to evaluate global scour and local scour and their effects on the laterally-loaded single piles under scoured conditions.

Mostafa [27] and Lin and Lin [21] also evaluated the scour effects on lateral behavior of pile groups in sands. Lin [11] and Lin et al. [16] performed an integrated analysis of pile-supported bridges under scoured conditions, which is reviewed at the end of this paper.

2 Behavior of Laterally-Loaded Piles Under Scoured Conditions

2.1 Overview

Behavior of laterally-loaded single piles in sands or clays has been well researched in the past through laboratory model tests, full-scale field tests, numerical analyses, and analytical solutions. The commonly used methods for evaluating the behavior of laterally-loaded single piles are mostly analytical or empirical. For example, Broms [4] developed analytical solutions and design charts for lateral load capacities of single piles in sand. In the practice, Reese's method [28] is commonly used to analyze laterally-loaded single piles in sand while Matlock's method [23] is used for laterally-loaded single piles in clay. To consider the pile group effect, Brown et al. [5] proposed the method of p-multiplier, which has been widely used in practice to evaluate the behavior of laterally-loaded pile groups. These methods may still be used to evaluate the behavior of laterally-loaded single piles and pile groups under global scour but need to be modified under local scour.

2.2 Single Piles

Effect of Global Scour. Diamantidis and Arnesen [7] found that when the scour bottom width was six times the pile diameter, local scour could be considered as global scour. Ismael and Han [9] conducted physical model tests with a scour width at least 6.3 times the pile diameter to evaluate the effect of global scour on the lateral load-displacement curves and lateral load capacities of single piles in dry sand. In their study, the pile was pinned at its bottom but it could have free rotation. Ismael and Han [9] found that the increase of the scour depth significantly reduced the lateral load capacities of the piles as shown in Fig. 2. The lateral load capacity ratio is defined as the ratio of the capacity with scour to that without scour. The calculated ratios were obtained using the design chart developed by Broms [4] without scour. The design chart considered the eccentricity of the lateral load relative to the ground surface. The increase of the scour depth is equivalent to the increase of the eccentricity and the reduction of the embedment of the pile. This excellent match indicates that the design chart developed by Broms [4] can be used to evaluate the lateral load capacities of piles due to global scour.

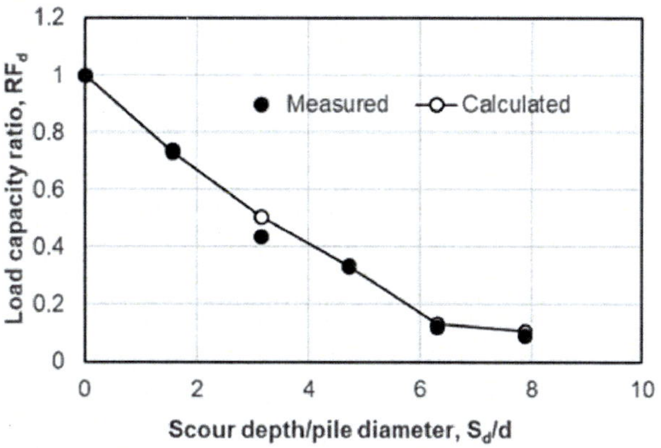

Fig. 2 Lateral load capacity ratio versus scour depth/pile diameter ratio for global scour (modified from Ismael and Han [9])

Ismael [8] also evaluated the bending moment change along the height of the pile with the increase of the scour depth at the failure load as shown in Fig. 3. The bending moment was normalized by the maximum moment in the pile without scour. Since the lateral load capacity of the pile decreased with the increase of the scour depth, the bending moment in the pile decreased with the increase of the scour depth. Figure 3 shows the maximum bending moment along the pile moved downward with the increase of the scour depth because the embedment depth of the pile decreased with the scour depth.

Removal of soil around piles changes stresses around the piles, which may affect the behavior of laterally-loaded piles. In the reduced-scale physical model tests, Ismael [8] did not simulate the process of soil removal around the pile; therefore, the effect of soil removal to cause possible stress history change cannot be evaluated. Lin et al. [17] theoretically investigated the effect of the stress history change of the

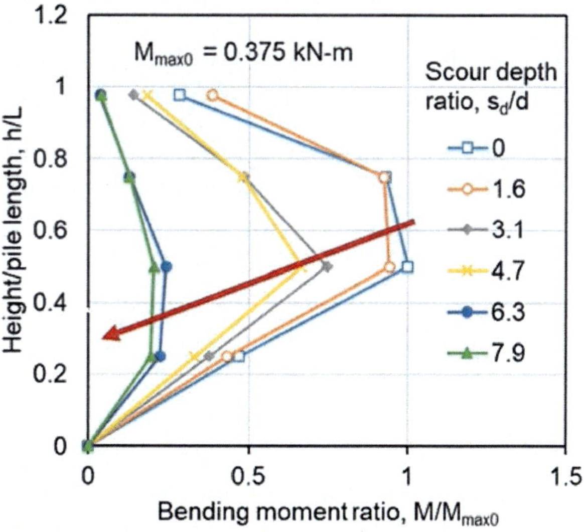

Fig. 3 Bending moment along the height of the pile at different scour depth (modified from Ismael [8])

remaining soil after global scour on the behavior of laterally-loaded piles in sands under a scoured condition while Lin et al. [14] investigated this effect on the behavior of laterally-loaded piles in clays. Lin et al. [14, 17] followed the concept of the CamClay model for soil rebound and the over-consolidation ratio and undrained shear strength relationship for the remaining soil. Lin et al. [14] found that global scour reduced the vertical stresses in the sand and increased its friction angle considering the stress-dependent friction angle and the over-consolidation ratio thus increasing the coefficient of lateral earth pressure at the interface between pile and sand. As a result, the lateral load capacity of the pile in the sand increased and its displacement decreased after considering the stress history effect. When piles were in clays, Lin et al. [17] found that global scour had more effect on the reduced effective stress than the increased over-consolidated ratio, thus reducing the undrained shear strength of the clay. As a result, the lateral load capacity of the pile in the clay decreased and its displacement increased after considering the stress history effect. In other words, ignoring the stress history effect due to global scour is conservative for the behavior of laterally-loaded piles in sands but un-conservative for that of piles in clays.

Lin et al. [19] investigated the global scour effect on the buckling loads of single piles fixed the base but having different pile head fixities. The soil support was modeled by multilinear stiffness springs in a structural software or by non-linear *p-y* curves in the Lpile software. These two methods resulted in similar lateral deflections under the same lateral loads. The analysis showed that the buckling loads decreased with the scour depth for all head fixities and the reduction of the buckling load was more significant at a smaller scour depth.

Effect of Local Scour. The effect of local scour on the behavior of laterally-loaded single piles may be considered as that of global scour to be conservative in practice. However, Lin et al. [13] found that considering local scour as global scour led to 49–68% larger groundline lateral displacements of single piles in sands under typical lateral loads. This difference implies that proper consideration of local scour is necessary and has been increasingly researched in recent years. Most recent studies on local scour around single piles have been based on circular piles. Square-shaped piles are also used in practice. Sheppard and Renna [29] suggested the use of equivalent or effective diameter based on the direction of water flow relative to the orientation of a pile. Ismael [8] conducted physical model tests to investigate the effect of scour-hole geometry and dimensions on the load capacities of laterally-loaded piles in sand as illustrated in Fig. 4 for some of the test setups as examples. Ismael [8] simulated the scour hole as a wedge cone with two symmetric scour slopes along the direction of lateral loading on the pile. In addition to the tests on global scour, these tests include four different scour bottom widths (0, 3.8d, 6.3d, and 10.5d), two scour slope angles (15° and 30°), and four scour depths (3.2d, 4.8d, 6.3d, and 7.9d) of scour holes. Details of these tests can be found in Ismael [8].

Based on the test results provided in Ismael [8], the effect of the slope angle on the lateral load capacity of the pile can be evaluated by calculating the ratio of the load capacity with a slope to that without a slope (i.e., global scour) at the same scour depth. Figure 5 shows that the scour slope increased the lateral load capacity of the

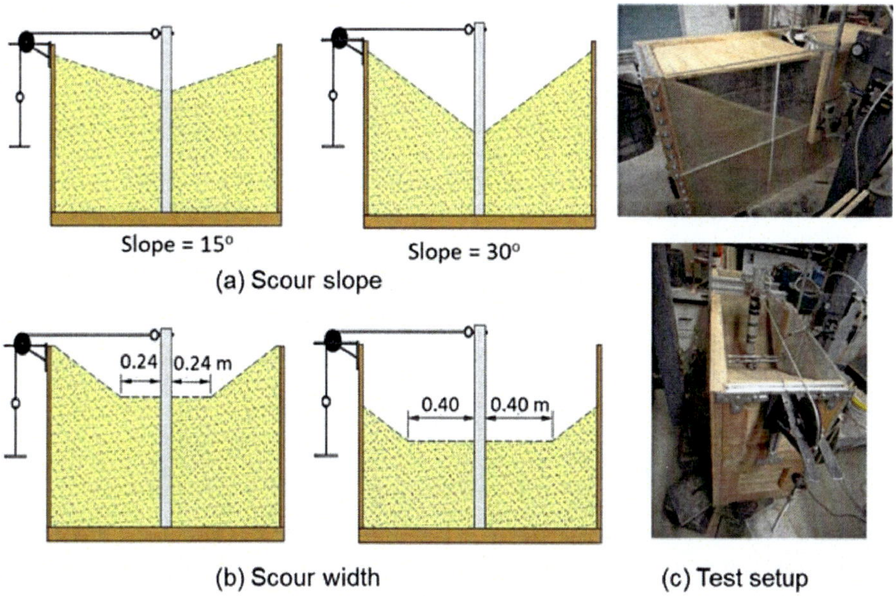

Fig. 4 Model tests for effects of scour-hole dimensions [8]

Fig. 5 Effect of scour slope on lateral load capacity ratio (generated from the data in Ismael [8])

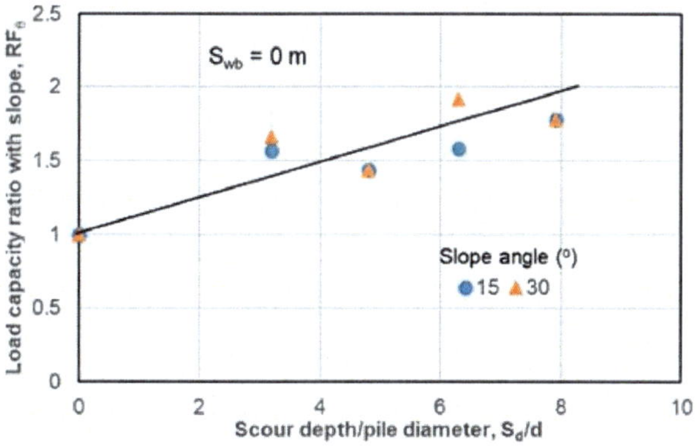

pile within the scour with zero scour width at the bottom; however, the slope angles of 15° and 30° did not make much difference. The increase of the scour depth shows more effect of the scour slope on the lateral load capacity because of the increase of the surcharge by the remaining soil slope.

Figure 6 shows the effect of the scour width at the bottom on the lateral load capacity ratio, defined as the ratio of the lateral load capacity of the pile at the scour width at the bottom to that due to global scour at the same depth. This ratio clearly shows that the increase of the scour width reduced the pile load capacity and the increase of the scour depth increased the effect of the scour width.

Ismael [8] also investigated the effect of scour depth on the lateral load capacity of the single pile under different scour slope and scour width, which is presented

Fig. 6 Effect of scour width at bottom on lateral load capacity ratio (generated from the data in Ismael [8])

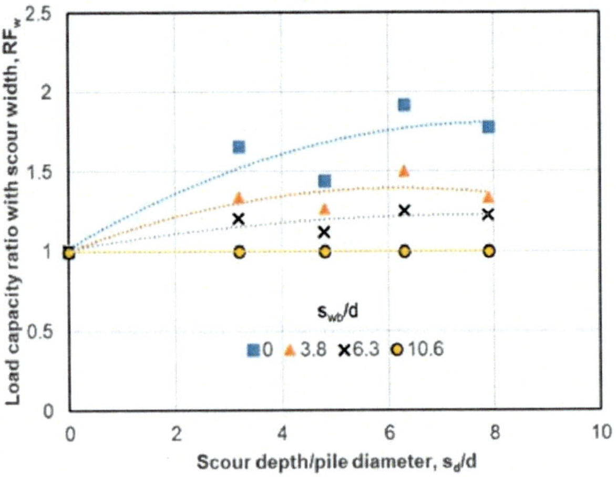

herein. This effect can be evaluated by combining the test results for global scour in Fig. 2 and the ratios of load capacities under local and global scour as presented in Figs. 5 and 6.

Lin et al. [13] proposed an analytical method to modify the failure mode of the soil wedge subjected to a lateral load from a pile in sand as proposed by Reese et al. [28] by considering a local scour hole in an inversed truncated cone shape as shown in Fig. 7. In this method, Lin et al. [13] proposed a concept of an equivalent soil wedge depth to the pile in the sand without scour based on an equal lateral load capacity to the pile with local scour. The local scour resulted in a reduced soil wedge depth so that the lateral load capacity of the pile decreased. Lin et al. [13] found that an increase of the scour depth significantly increased the pile lateral displacement and

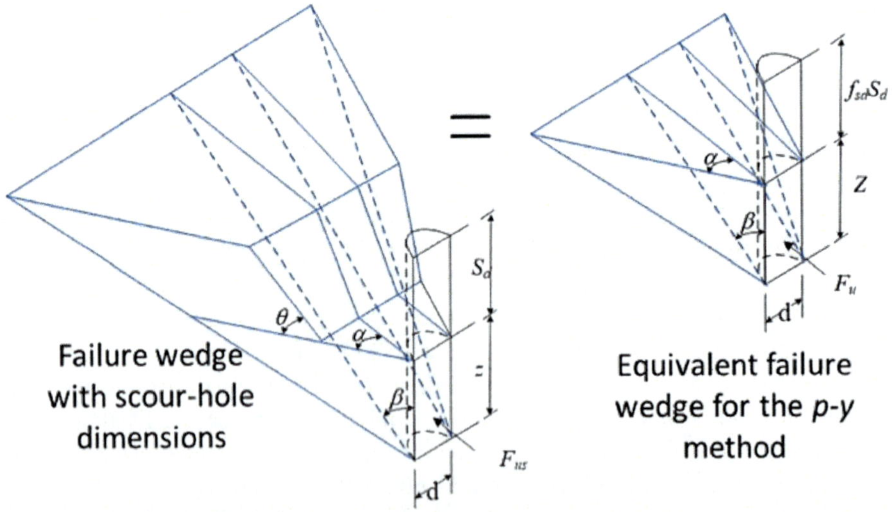

Fig. 7 Modified failure mode for a laterally-loaded pile in sand with local scour [13]

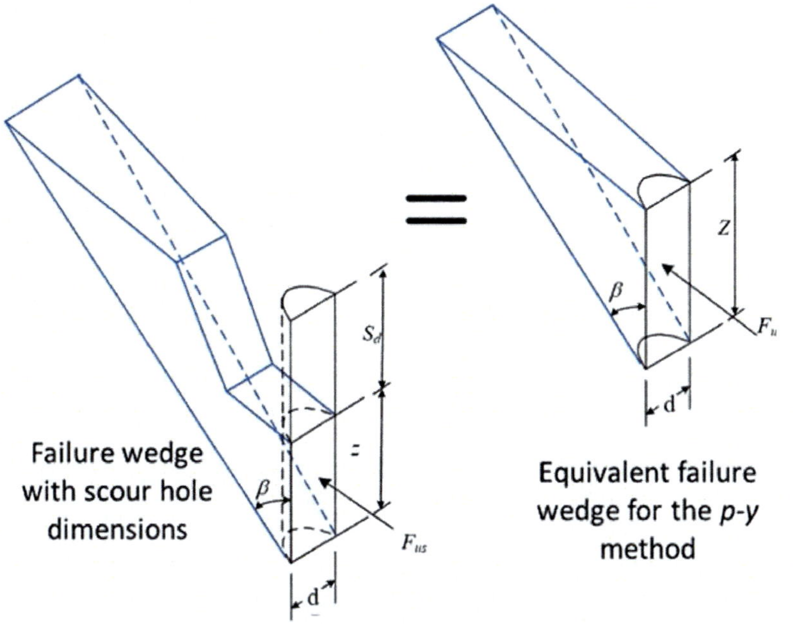

Fig. 8 Modified failure mode for a laterally-loaded pile in clay with local scour [12]

the maximum bending moment but the scour width and the scour-hole slope angle had relatively less effects, especially when $S_{wb} > 8d$. In addition, Lin and Lin [20] found that local scour reduced lateral load capacities of piles in dense sand by 10% more than those in loose sand.

Lin et al. [12] proposed another analytical method to modify the failure mode of the soil wedge subjected to a lateral load from a pile in clay as proposed by Matlock [23] by considering a local scour hole in an inversed truncated cone shape as shown in Fig. 8. In this method, Lin et al. [12] also adopted the concept of an equivalent soil wedge depth to the pile in the clay without scour based on an equal lateral load capacity to the pile with local scour. The difference of this method from that for the pile in the sand is the angle of the distributed wedge. For the clay, this angle is equal to zero. The local scour also resulted in a reduced soil wedge depth so that the lateral load capacity of the pile decreased. These two analytical models were verified by the three-dimensional numerical analyzes conducted by Lin et al. [12, 13]. It should be pointed out that they did not consider the effect of stress history change due to global and local scour.

Similar to the experimental study on global scour, Ismael [8] did not simulate the process of soil removal around the pile due to local scour; therefore, the effect of soil removal to cause possible stress history change cannot be evaluated either. Zhang et al. [32] theoretically investigated the effect of the stress history change of the remaining soil after local scour on the behavior of laterally-loaded piles in clays. Zhang et al. [32] simulated the local scour hole in an inversed truncated cone

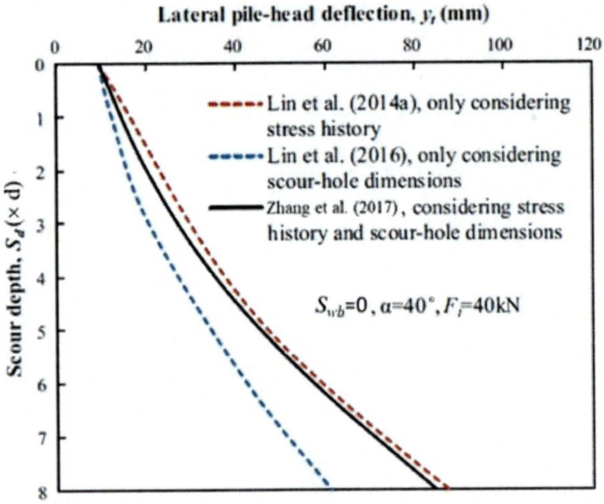

Fig. 9 Effect of stress history and scour-hole dimensions [32]

shape and used Mindlin Green's function for vertical and horizontal loads in a semi-infinite half-space to calculate the scour-induced stress reduction in the soil. As a result, the over-consolidated ratio and undrained shear strength of the clay after local scour were estimated using the same concept of the CamClay model for soil rebound and the over-consolidation ratio and undrained shear strength relationship for the remaining soil as used by Lin et al. [14]. By considering the reduction of the embedment length of the pile and the reduced undrained shear strength of the clay due to local scour, Zhang et al. [32] used the Matlock method [23] to predict the p-y curve of the pile due to local scour. Figure 9 shows the comparison of the calculated results considering stress history only, considering scour-hole dimensions only, and both stress history and scour-hole dimensions, indicating the stress history had more effect on the behavior of laterally-loaded single piles in the clay than the scour-hole dimensions in this analysis.

Lin and Wu [22] evaluated different calculation models for vertical stresses after local scour including the US Federal Highway Administration (FHWA) method for driven piles (FHWA-DP), the FHWA method for drilled shafts (FHWA-DS), the American Petroleum Institute method (API), the improved method proposed by Lin and Wu [22], and the simplified method proposed by Lin and Wu [22] as shown in Fig. 10. The FHWA-DP method did not consider the local scour effect while the FHWA-DS method considered a linear reduction of the vertical stresses from the bottom of the scour hole to the depth of 1.5 times the scour depth. The API method considered a linear reduction of the vertical stresses from the bottom of the scour hole to the depth of six times the pile diameter. Lin and Wu [22] proposed an improved method to estimate the reduced vertical stresses along the center of the pile at different depths by using the Boussinesq solution for an embankment type of loading. Based on the parametric study, Lin and Wu [22] found that their improved method could be simplified by setting the influence depth equal to 3.5 times the scour depth. After the calculation of the reduced vertical stress at a specific depth below the

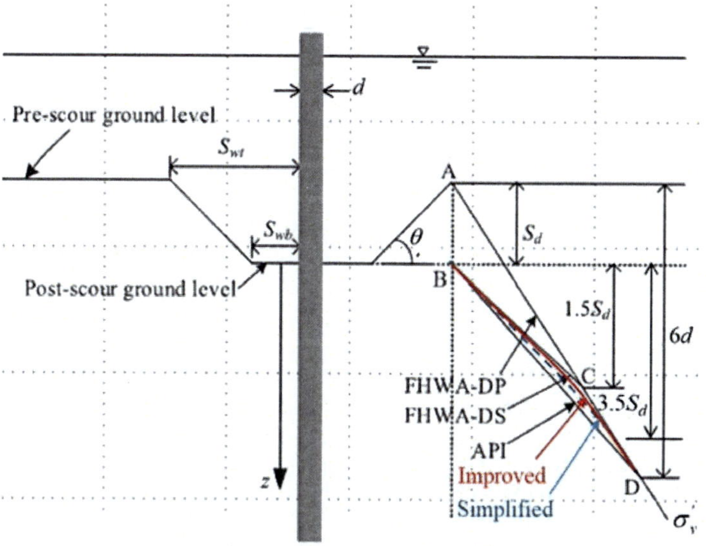

Fig. 10 Reduced vertical stress calculation models after local scour (modified from Lin and Wu [22])

bottom of the scour hole, the p-y curve at this depth can be calculated using available methods for laterally-loaded piles in soil without scour by the reduced vertical stress replacing the overburden stress and the equivalent depth depending on whether it is above or below the influence depth. Lin and Wu [22] examined the differences of the calculated lateral load capacities of single piles by these different methods as presented in Fig. 11. The lateral load capacity ratio was defined as the ratio of the lateral load capacity of the pile after scour to that before scour. Figure 11 shows that the API method, the improved method, and the simplified method calculated similar lateral load capacity ratios; however, both FHWA methods over-predicted the lateral load capacity ratios. Instead of the calculated reduced vertical stress, Lin and Lin [20] suggested to convert the vertical stress into an equivalent depth by dividing the vertical stress by the soil effective unit weight. This equivalent depth can be used in the *p-y* curve solution by Reese et al. [28] as Lin et al. [13] did to consider the local scour effect.

2.3 Group Piles

Behavior of laterally-loaded group piles in sands or clays has also been well researched in the past through laboratory model tests, full-scale field tests, numerical analyzes, and analytical solutions. Under lateral loads, an individual pile in a pile group often behaves weaker and softer than a single pile due to overlapping of stresses in soil between neighboring piles (i.e., pile-soil-pile interaction). To consider

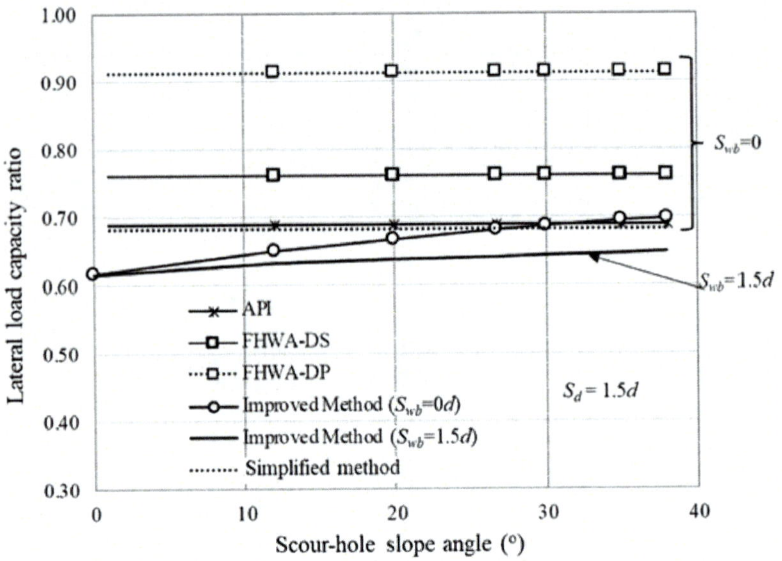

Fig. 11 Effect of calculated vertical stresses after local scour by different methods on the lateral capacity ratio (modified from Lin and Wu [22])

the pile group effect, the method of p-multiplier proposed by Brown et al. [5] has been widely used in practice to evaluate the behavior of a laterally-loaded pile group. The p-multiplier method includes applying a reduction factor, i.e., so-called a p-multiplier (f_m) to the *p-y* curve of a single pile to generate a modified *p-y* curve of a corresponding individual pile in the pile group under lateral loads. The p-multiplier depends on pile spacing, pile relative location (front, middle, rear, edge, or corner) in the pile group, soil properties, and pile head fixity conditions. The sum of the *p-y* curves for all piles in the group results in a *p-y* curve for the pile group.

Scour around group piles is more complex than that around single piles. There may be three types of scour: (1) global scour, (2) local scour around a pile group (i.e., group local scour), and (3) local scour around an individual pile (i.e., pile local scour). Based on the laboratory tests conducted for six pile groups of different pile spacing, Sumer et al. [30] found pile local scour holes developed around individual piles inside a group local scour hole around the largely-spaced pile group (pile spacing $s = 5d$). For the closely-spaced pile groups ($s \leq 3d$), however, pile local scour holes overlapped with each other. Since individual piles within a pile group are typically spaced at $3d$ or $4d$ in most practices, pile local scour can be ignored and a single group scour hole can be used to represent the local scour around a pile group.

To estimate group local scour depths, Sheppard and Renna [29] suggested three different cases before scour: (1) pile caps above riverbed (i.e., group piles exposed), (2) pile caps partially exposed, and (3) pile caps completely buried). These conditions affect scour-hole geometry and dimensions including scour depths. Sheppard and Renna [29] provided detailed procedures for estimating scour depths for these cases. Mostafa and Agamy [27] conducted an experimental study to evaluate scour-hole geometry and dimensions around the pile groups with two piles in a side-by-side arrangement, two piles in a tandem arrangement, and three piles in a triangular arrangement versus those around single piles. Based on the reported photos, the scour-hole shape for two piles was approximately elliptical while that for three piles was approximately circular. They found that the scour depth for the case with the pile group was generally greater than that for the case with a single pile depending on the group configuration and the gaps between piles. Amini et al. [1] showed that scour holes generated around 3 × 5 pile groups during the flume tests had a rounded square shape. Lin and Lin [21] pointed out that group piles under scoured conditions have two group effects: (1) group effect due to overlapped stresses and (2) group effect due to increased scour depths by increased flow velocity and turbulence between piles. As a result, the load capacity of an individual pile in the pile group is lower than that of a single pile.

Effect of Global Scour. The double group effect on the lateral load capacity of an individual pile in the pile group may be evaluated by two group reduction factors or efficiencies as shown in Eq. (1):

$$F_{gs} = F_1 f_m f_s \tag{1}$$

where F_{gs} = the lateral load capacity of an individual pile in the pile group under global scour, F_1 = the lateral load capacity of a single pile without scour, f_m = the p-multiplier for the group effect due to stress overlapping in pile-soil-pile interaction, and f_s = the group reduction factor due to global sour.

Mostafa [26] used the software program GROUP V.7.0 to generate the lateral load-head displacement curves for laterally-loaded single piles and group piles in a side-by-side arrangement or a tandem arrangement with and without global scour and investigated the effects of scour depth, pile spacing, pile arrangement, and pile slenderness ratio. As expected, an increase in the scour depth and/or a decrease of the pile spacing reduced the lateral load capacity of the pile group. Mostafa [26] also showed that pile groups in the side-by-side arrangement had larger displacements and bending moments as compared with single piles and pile groups in a tandem arrangement due to the combined effect of scour depth and pile-soil-pile interaction and scour had more effect on the lateral group load capacity and displacement than the pile-soil-pile interaction. Furthermore, Mostafa [26] concluded that shorter piles are more vulnerable to global scour and the effect of the pile slenderness on the reduction in the lateral pile group capacity became insignificant after the pile slenderness ratio was greater than approximately 12.5.

Effect of Group Local Scour. Lin and Lin [21] extended their approach developed by Lin and Lin [20] for single piles in sand under pile local scour to analyze laterally-loaded pile groups in sand with group local scour. In this extended method, they estimated the reduced vertical stress for the central pile using the Boussinesq solution for the condition after scour and then converted the reduced vertical stress into an equivalent depth, which is used for the *p-y* curve analysis. This method adopted the procedure recommended by Sheppard and Renna [29] to estimate the equivalent or effective diameter of a pile group based on the direction of water flow relative to the orientation of a pile. Lin and Lin [21] made two important assumptions: (1) the p-multiplier (f_m) for each pile is unchanged by scour and (2) the bottom scour width around the pile group is zero. The first assumption was verified by numerical results. Their results showed that treating the group local scour as a global scour by removing the whole scoured soil resulted in lower lateral group load capacities, indicating the benefit of considering the geometry of a group local scour hole in practice. However, limited studies have been conducted so far considering the geometry of the group local scour hole for pile groups, clearly, future research is needed to develop more general solutions for evaluating the behavior of laterally-loaded pile groups in sands and clays.

3 Lateral Behavior of Pile-Supported Bridges Under Scoured Conditions

3.1 Overview

The above sections discusses the behavior of laterally-loaded single piles and pile groups in sands or clays under scoured conditions. In actual applications, single piles and/or pile groups are only foundations to support bridges. The behavior of pile foundations may affect the behavior of pile-supported bridges. However, interactions among pile foundations, bridge structures, and abutments contribute to the performance of pile-supported bridges. As discussed earlier, global scour and local scour affect the behavior of laterally-loaded single piles and pile groups in sands and clays under scoured conditions. It is expected that they should affect the lateral behavior of pile-supported bridges under scoured conditions.

3.2 Integrated Analysis for Pile-Supported Bridges Under Scoured Conditions

To evaluate the influence of laterally-loaded piles or pile groups under scoured conditions on the performance of pile-supported bridges, integrated analysis of interactions among pile foundations, bridge structures, and abutments is necessary. Lin [11] and Lin et al. [16] developed a structural analysis model for a pile-supported bridge under a scoured condition as shown in Fig. 12. In this model, a structural software was used to model the bridge girders, abutment supports, piers, pile caps, and group piles with soil supports as springs. These springs were placed at different depths of each pile and had non-linear (i.e., multilinear) reaction-displacement behavior as described by a *p-y* curve. The scour around each pile was modeled by removing springs according to the scour depth. The *p-y* curve for each remaining spring was

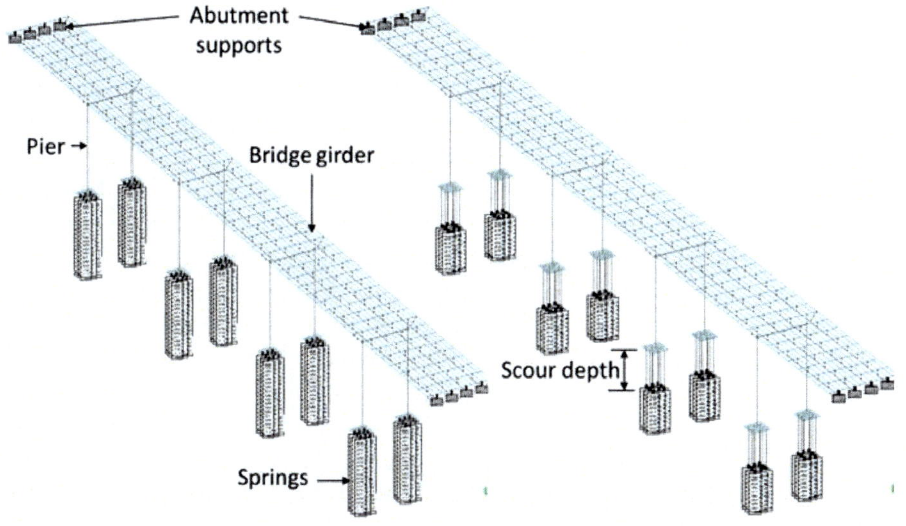

Fig. 12 Integrated analysis model for a pile-supported bridge under a scoured condition (modified from Lin [11])

determined by a calculation module developed by Lin [11] to account for the scour effect. Lin et al. [16] considered global scour but ignored the stress history effect while Lin [11] considered both the global scour and the stress history effect. The analyses in both studies considered vertical loads (dead and live loads), water loads, wind loads, and debris loads. Both analyses showed that an increase of the scour depth significantly increased the upstream and downstream pile group deflections but slightly increased the abutment deflections. At the same time, the increase of the scour depth significantly increased the shear forces at the abutments but slightly increased the shear forces at the pile caps. Lin [11] showed that the stress history change of the remaining soil had less effect on the behavior of the pile groups under the bridge than that without a bridge. This is because the girders and the abutments restrained the pile group movement in the bridge system. Lin et al. [16] also showed that the increase of the scour depth reduced the buckling load of the individual pile in the pile group and the reduction was the most for the upstream pile, followed by the middle and downstream pile due to the water loads, the wind loads, and the debris loads applied from the upstream to the downstream.

4 Conclusions

This paper reviewed the recent studies on the behavior of laterally-loaded piles under scoured conditions at bridges. The effects of global scour and local scour on the behavior of laterally-loaded single piles and pile groups were examined. Removal of soil by scour changes the stress history of the remaining soil, which increases the strengths of the remaining sand but reduces the strength of the remaining clay. The geometry and dimensions of a local scour hole affect the lateral load capacity of a single pile or a pile group. The scour depth has more influence on the behavior of laterally-loaded piles than the scour-hole width and slope angle. The scour-hole effect can be evaluated by a reduced equivalent soil depth based on the reduced wedge lateral resistance or the reduced vertical stress, which is used to modify the p-y curve of the pile in the soil under a scoured condition. Individual piles in a pile group under a scoured condition have double group effects due to overlapped stresses (pile-soil-pile interaction) and increased scour depths so that the group lateral load capacity is more significantly reduced. Scour also reduces the buckling load of piles in soils. The effect of scour on the performance of pile-supported bridges should be evaluated by an integrated analysis that considers hydraulic, geotechnical, and hydraulic aspects. The scour effect in the integrated analysis can be considered by removing soil springs around piles based on the scour depth and adjusting the spring stiffness in terms of the modified p-y curve accounting for the stress history change and the scour-hole dimensions. Further research is needed to better evaluate the behavior of laterally-loaded pile groups under group local scour and pile local scour.

Acknowledgments The author would like to acknowledge the assistance received from Dr. Cheng Lin at the University of Victoria in Canada and Dr. Fayun Liang at Tongji University in China during the preparation of this paper.

References

1. Amini A, Melville BW, Ali TM, Ghazali AH (2012) Clear-water local scour around pile groups in shallow-water flow. J Hydraul Eng 138(2):177–185
2. Arneson LA, Zevenbergen LW, Lagasse PF, Clopper PE (2012) Evaluating scour at bridges, 5th edn., FHWA-HIF-12-003 (HEC 18), Federal Highway Administration, Washington, D.C.
3. Briaud JL (2010) Bridge foundation scour. Geotech Eng J SEAGS AGSSEA 41(2)
4. Broms BB (1964) Lateral resistance of piles in cohesionless soils. ASCE J Soil Mech Found Div 90(SM3, Part 1):123–156
5. Brown DA, Morrison C, Reese LC (1988) Laterally loaded behavior of pile group in sand. ASCE J Geotech Eng 114(11):1261–1276
6. Butch GK (1996) Scour-hole dimensions at selected bridge piers in New York. In: Bathala C (ed) North American Water and Environment Congress & destructive water. American Society of Civil Engineers, New York, pp 3043–3051
7. Diamantidis D, Arnesen K (1986) Scour effects in piles structures—a sensitivity analysis. Ocean Eng 13(5):497–502
8. Ismael O (2014) Evaluating the behavior of laterally loaded piles under a scoured condition by model tests. Master's thesis, The University of Kansas, Lawrence, Kansas, USA
9. Ismael OK, Han J (2015) Model tests of laterally loaded piles under a horizontally scoured condition. In: The International Foundations Congress and Equipment Expo 2015, San Antonio, Texas, March 17–21
10. Li F, Han J, Lin C (2013) Effect of scour on the behavior of laterally loaded single piles in marine clay. J Mar Georesour Geotechnol 31:271–289

11. Lin C (2012) Evaluation of lateral behavior of pile-supported bridges under scoured conditions. Ph.D. dissertation, The University of Kansas
12. Lin C, Han J, Bennett CR, Parsons RL (2016) Analysis of laterally loaded piles in soft clay considering scour-hole dimensions. Ocean Eng 111:461–470
13. Lin C, Han J, Bennett CR, Parsons RL (2014) Analysis of laterally-loaded piles in sand considering scour hole dimensions. ASCE J Geotech Geoenviron Eng 04014024-1–13
14. Lin C, Han J, Bennett CR, Parsons RL (2014) Behavior of laterally-loaded piles under scour conditions considering stress history of undrained soft clay. Technical note. ASCE J Geotech Geoenviron Eng 06014005-1–6
15. Lin C, Bennett CR, Han J, Parsons RL (2014) Effect of soil stress history on scour evaluation of pile-supported bridges. ASCE J Performance Constructed Facil 04014178-1–11
16. Lin C, Bennett CR, Han J, Parsons RL (2012) Integrated analysis of the performance of pile-supported bridges under scoured conditions. Eng Struct 36:27–38
17. Lin C, Bennett CR, Han J, Parsons RL (2010) Scour effects on the response of laterally loaded piles considering stress history of sand. Comput Geotech 37:1008–1014
18. Lin C, Han J, Bennett C, Parsons RL (2014) Case history analysis of bridge failures due to scour. In: International symposium of climatic effects on pavement and geotechnical infrastructure, ASCE, pp 204–216
19. Lin C, Bennett C, Han J, Parsons RL (2010) p-y based approach for buckling analysis of axially loaded piles underscoured conditions. In: Proceedings of SEI 2010 structures congress, Orlando, FL
20. Lin YJ, Lin C (2019) Effects of scour-hole dimensions on lateral behavior of piles in sands. Comput Geotech 111:30–41
21. Lin Y, Lin C (2020) Scour effects on lateral behavior of pile groups in sands. Ocean Eng 208:107420
22. Lin C, Wu R (2019) Evaluation of vertical effective stress and pile lateral capacities considering scour-hole dimensions. Can Geotech J 56(1):135–143
23. Matlock H (1970) Correlation for design of laterally loaded piles in soft clay. In: Proceedings of the II annual offshore technology conference, Houston, TX, pp 277–594
24. Melville B (2008) The physics of local scour at bridge piers. In: Proceedings of the fourth international conference on scour and erosion
25. Mostafa YE (2012) Effect of Local and global scour on lateral response of single piles in different soil conditions. Engineering 4:297–306
26. Mostafa YE (2012) Design considerations for pile groups supporting marine structures with respect to scour. Engineering 4:833–842
27. Mostafa YE, Agamy AF (2011) Scour around single pile and pile groups subjected to waves and currents. Int J Eng Sci Technol 3(11):8160–8178
28. Reese LC, Cox WR, Koop FD (1974) Analysis of laterally loaded piles in sand. In: Proceedings of the VI annual offshore technology conference, Houston, TX, pp 473–485
29. Sheppard DM, Renna R (2010) Bridge scour manual. Florida Department of Transportation
30. Sumer BM, Bundgaard K, Fredsøe J (2005) Global and local scour at pile groups. Int J Offshore and Polar Eng 15(3):204–209
31. Wang C, Yu X, Liang FY (2017) A review of bridge scour: mechanism, estimation, monitoring and countermeasures. Nat Hazards 87(3):1881–1906
32. Zhang H, Chen SL, Liang FY (2017) Effects of scour-hole dimensions and soil stress history on the behavior of laterally loaded piles in soft clay under scour conditions. Comput Geotech 84:198–209

Sustainable Hard and Soft Measures for Coastal Protection

V. Sundar

Abstract The coastal regions across the globe are more densely populated; this effect could be endorsed to urban development including the infrastructure for ports, inland waterways, trade and commerce, tourism, fisheries, hub for power plants, etc. Owing to the growing population demands many residential, industrial and commercial units have been constructed in the close vicinity/proximity of the shoreline. The developmental activities across the shore mandate the construction of facilities that promote industrial growth such as fishing harbours, ports, jetties, training walls, quays and wharfs as well as facilities for tourism development such as surfing friendly seabed slopes, added beach width for recreational activities and establishment of beach resorts, quays and wharfs. Predominant coastal hazards such as coastal flooding, coastal erosion, sea level rise, storm surge and seawater inundation are common threats to sensitive coastal communities. Alongside the prevailing natural causes of melting glaciers due to global warming and sea level rise, extreme events such as storms and cyclones; man-made artificial structures in the near shore regions/in between shallow and intermediate water depths can potentially destabilize the shoreline and lead to excessive coastal erosion if not addressed carefully. This paper details the vast experience gained in the field of coastal protection measures along the coastline of India.

Keywords Coastal erosion · Sustainable coastal protection · Hard engineering measures · Soft engineering measures

1 Introduction

The coastal stretches are highly dynamic in nature which is attributed to varying forces exerted upon them due to waves, tides, currents, suspended and bed load sediments, etc. In addition, the seasonal variations also bring about cyclones, storm

V. Sundar (✉)
Department of Ocean Engineering, Indian Institute of Technology Madras, Chennai, India
e-mail: vsundar@iitm.ac.in

C. N. V. S. Reddy and S. Sassa (eds.), *Scour and Erosion Related Issues*,
Lecture Notes in Civil Engineering 177,
https://doi.org/10.1007/978-981-16-4783-3_4

surges, depressions, tsunami and tidal bores. The varying combination of the aforesaid forces tends to disrupt the equilibrium of the coastline. All the coastal stretches are not subjected to extreme forces, some of the stretches are naturally sheltered in a bay-like area where the regular impact of waves, tides and currents are relatively less and are typically exempted from any form of coastal hazard whereas, a significant stretch of the Indian coastline is exposed to the extreme wave climate leading to coastal erosion.

The artificial or man-made causes for coastal erosion are the construction of conventional coastal structures such as breakwaters, jetties, fishing harbours, ports to promote trade and fisheries. This can be either long-term or short-term effect on the coastline adjacent to these structures which lead to shoreline oscillation due to the interception of the long-shore littoral drift currents within the surf zone which redistributes the sediment supply on the downdrift side causing erosion and loss of beach width termed as erosion. The confluence of rivers and oceans may also contribute to erosion on its downdrift side, whereas, on the updrift of the structures or river mouths are characterized by beaches or sand bar formation, which also possess challenges in maintaining the river mouth connected to the Sea. Natural causes also play a crucial role in aggravating coastal hazards, the seasonal storms and cyclones significantly increase the near shore water depths resulting in seawater flooding and erosion. Extreme events such as the tsunami result in huge damage to life and property, wreck the existing coastal infrastructure, causes extensive coastal erosion during run-down, etc.

To regulate and stabilize the coastline, shore protection measures are in place to win back the lost sediments due to erosion. In most of the scenarios, where a short-term effect in loss of sediments, the "do nothing" approach would naturally restore equilibrium. If the causes resulting in coastal erosion are adverse, to protect, to accommodate, or to retreat approaches (Fig. 1) can be attempted. This paper would discuss a few case studies on the coastal protection aspects.

The Coastal Regulation Zone (CRZ) notification was issued in February 1991 to regulate the activities in the coastal area by the Ministry of Environment and Forests (MoEF), Govt of India, which states that, "the coastal land up to 500 m from the High Tide Line (HTL) and a stage of 100 m along banks of creeks, lagoons, estuaries, backwater and rivers subject to tidal fluctuations, is called the CRZ". The CRZ's includes only the inter-tidal zone and land part of the coastal area and does not include the ocean part. Most of the planned protection measures are located within the CRZ's and sometimes probing into the ocean, since they are in a highly sensitive zone utmost care and attention to detail for the design is given by planners, engineers and Govt. agencies prior to implementation of coastal protection measures. All the existing options are explored to present an ecological and economically effective solution to redress the coastal erosion.

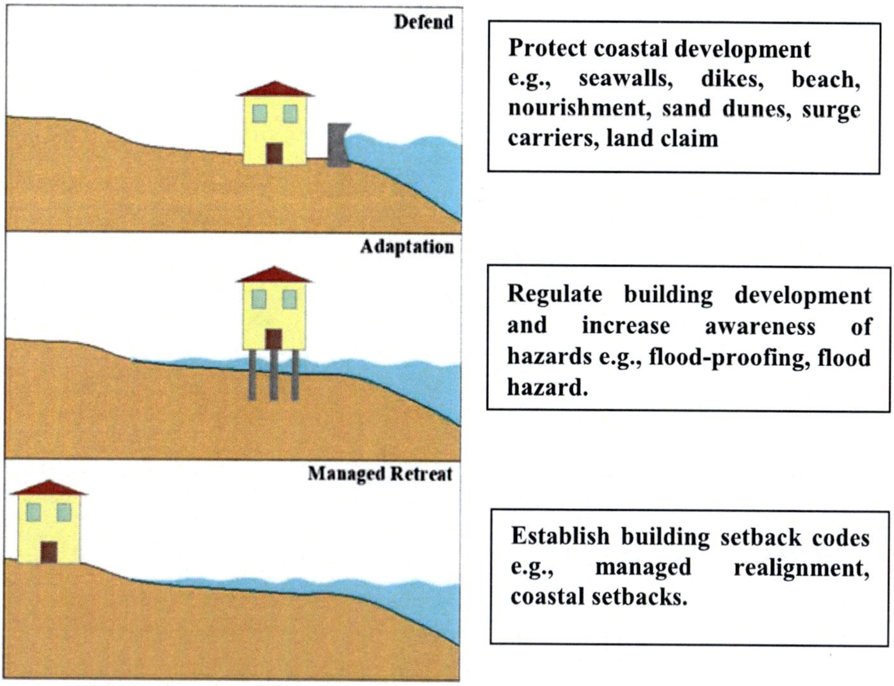

Fig. 1 Basic approaches to handle coastal erosion

2 Why Is Coastal Protection Necessary?

The Indian sub-continent has a vast area of about 3.287 million km^2 with a coastline length of about 7500 km (including mainland and island) which is ranking only next to China in terms of the extent of the population affected due to climate change, subsidence and socio-economic reasons as pointed out by Nicholls et al. [8] that is projected in Fig. 2. It is easier to adapt the "do nothing" approach and choose the "retreat" option and relocate the near coast settlements further landward. Although, it is a simpler plan it is not economically viable to relocate the densely populated coastal stretches. The coastal metropolitan cities are a hub for all kinds of trade related activities, major nuclear and thermal power plants are located near the ocean for ease in the disposal of contaminant water and wastes, fisheries and salt mining are amongst the important coastal industries. Moreover, many culturally significant ancient structures are built along the shore with immense tourism potential which cannot be sacrificed to the sea. The proper planning, design, execution and maintenance of coastal protection measures could be economically as well as ecologically viable.

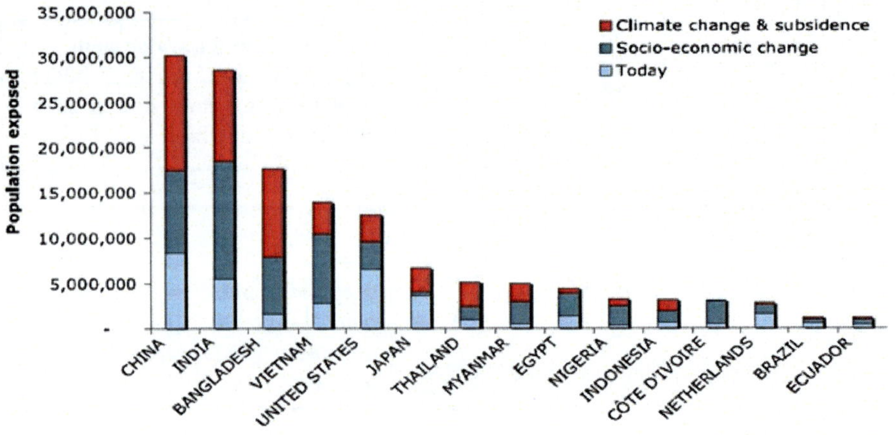

Fig. 2 Ranking of countries affected by climate change [8]

3 Preliminary Studies

The primary step is to identify the causes for erosion, whether it is temporary or persistent. This can be identified through field studies or satellite imaging. Secondly, a detailed field study and data collection regarding the nature of the coast, soil properties, sediment concentration, physical features, bathymetry, tide cycles, wave climate and frequency of extreme events need to be assessed. Lastly, the socio-economic factors including the aesthetic appeal of the coast, safety evaluation, public convenience, etc. need to be worked out, alternative temporary and more economic options are to be investigated and modelled with the aid of numerical or physical model studies in the laboratories. In the event of a long coast, for instance, that of the entire maritime state, a preliminary survey followed by identification of vulnerable stretches and evolving with conceptual designs need to be prepared. From the conceptual designs, detailed design needs to be done with the recent shoreline positions and bathymetry. If the necessary funds are available to execute the planned coastal protection measure, it can be commissioned under suitable weather conditions to arrive at optimum results. As the different seasons (South West, North East and Non-monsoon) dictate the direction the sediment transport or littoral currents, the construction sequence must be planned ahead. This is crucial at a locations where the groin field is planned as a coastal protection measure.

4 Coastal Protection Measures

4.1 General

The shore protection measures adopted can either be classical methods such as construction of groin field, seawall, revetments, or novel methods such as employing vegetation cover, geo-synthetic materials, beach nourishment or dune stabilization. The choice of adopting an established shore protection measure for a specific field problem needs to be done after scrutinizing the local field conditions and ensuring that the proposed measure is sustainable up to or beyond its design life. The coastal protection measures, in general, as projected in Fig. 3 is broadly classified as hard measures (groins, seawalls, offshore detached breakwaters) and soft measures (Artificial Beach Nourishment, Bio-Shields/Vegetation, dune stabilization, application of geo-synthetics). Erstwhile a combination of both can also be implemented.

The **hard** measures are conventional methods of coastal protection that consist of gravity type structures usually made up of rubble mound or precast concrete units. Implementing these measures yield immediate results, yet they are expensive,

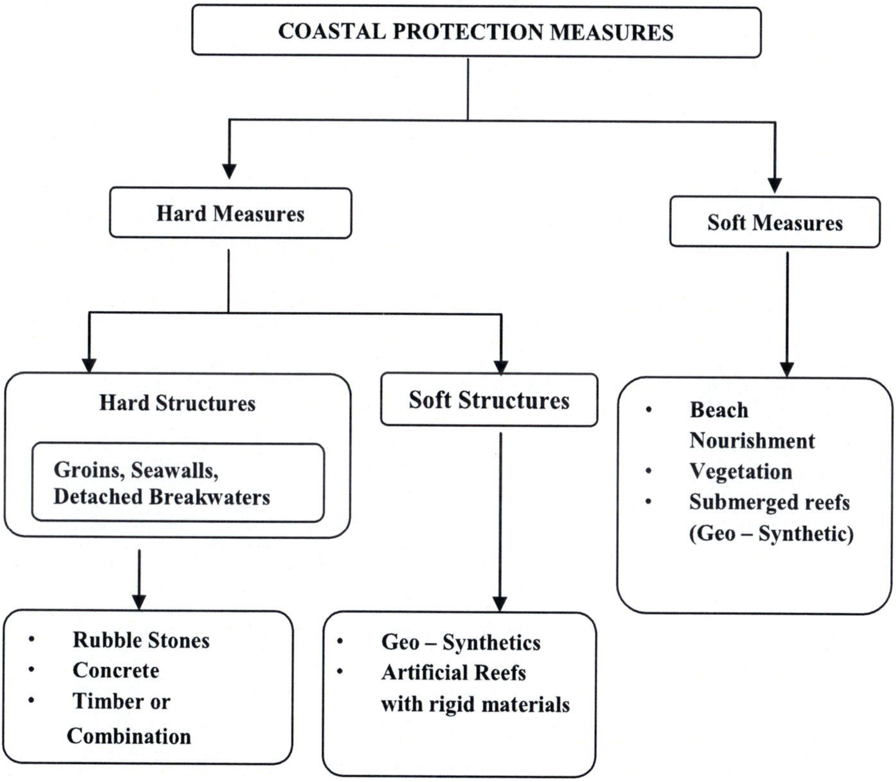

Fig. 3 Broad classification of coastal protection measures

unaesthetic, leave behind an ecological footprint and sometimes difficult to construct, however, its effects might be adverse if not properly planned and executed and are irreversible. However, they are a well established and the most reliable method to attain shoreline stability within a short possible time span. A few examples of successful implementation and performance of hard engineering measures are discussed in brief in the following sections. Adopting natural rocks for the armour layers for the different hard measures, has in fact resulted in its depletion forcing precast concrete blocks. Several types of concrete armour blocks (such as tetrapods, accropodes, dolos, x-blocks, etc.) are available which can be laid as single or double layers, the details of which are discussed in the coastal Engineering Manual [3]. Chandramohan et al. [2] modified the stem length of the DOLOS, termed it as KOLOS that was adopted for the breakwaters of the port at Krishnapatnam, Andhra Pradesh, India.

The **soft** engineering measures form the other contrast spectrum of coastal protection techniques, which leaves behind a minimum or almost no ecological footprint. In comparison with the hard structures, the shoreline response to these measures is rather a slow process. The popular soft engineering measures are artificial beach nourishment, dune stabilization and vegetal cover protection. These methods require a higher degree of expertise for planning and execution in the field. The recent developments in the application of geo-synthetics in coastal engineering problems can be termed as soft engineering structures. Pilarczyk [12] and Koerner [6] have discussed in detail the application and design guidelines of geo-synthetic materials across various civil, hydraulic and coastal engineering perspectives. Although geo-textile was employed for road construction in South Carolina, it was Recio and Oumeraci [13, 14] who pointed out the usefulness of the application geo-synthetic materials for marine works. The conventional structures such as seawalls, submerged breakwaters, reefs, etc. can be entirely or partially made from geo-synthetic materials. This is a viable option for remote locations where the transport of heavy boulders and concrete is difficult.

4.2 Hard Engineering Measures

Groin Field

Groins are shore connected structures extending from the shore and probing into the ocean up to or beyond the surf zone width and influence the local morpho-dynamics of the coasts by altering the sediment distribution and trapping sediments between groins. They can be designed in varying shapes and lengths depending on the site of implementation. The coastal city of Chennai houses the world's second-longest beach the Marina beach immediately south of its major port (initial breakwaters of the port was constructed in 1881), which intercepts the long shore sediment transport resulted in accretion on the updrift side (south of the breakwaters) and erosion on its downdrift side, i.e., on the north of the breakwaters. The images of the eroded north Chennai coastline are projected in Fig. 4a. This effect is attributed to the predominant

Fig. 4 **a** Eroded stretches of North Chennai coastline (before 2004). **b** Accreted stretches of North Chennai coastline (after 2004)

net littoral transport rate directed towards the north. Several temporary measures to protect the coast failed miserably since the coast experiences an annual northerly net long shore sediment transport to an extent of 0.8 million m^3. A proposal was made by Sundar et al. [18] to construct 10 number of transitional groins north of Chennai port (from 13° 08′ 48.36″ N, 80° 18′ 50.10″ E to 13° 10′ 47.80″ N, 80° 19′ 18.87″ E) to resolve the five decades long erosion problem. This protection measure was observed to be effective in restoring the lost beach width (Fig. 4b) which performed satisfactorily well even during the Indian Ocean Tsunami as reported in the post-tsunami survey of Tamilnadu coast by Sundar [19]. A detailed study by Sanjeevi and Jayaprakash [15] reported that about 94,465 m^2 of land has been gained in the Royapuram area and 35,660 m^2 in Ramakrishna Nagar area and on an average, 68,757 m^2 of land has been regained between 2004 and 2011, post the implementation of the groin field. An aerial photograph of the shoreline in mid-2020 is projected in Fig. 5, which highlights the success of the groin field, 16 years after implementation. About 25 stretches of the Kerala coast experiencing perennial erosion were protected

Fig. 5 Aerial view of North Chennai coastline (May 2020)

by well-planned groin fields through comprehensive scientific methods of evaluation, which has yielded positive results of winning back the lost beaches as detailed by Sundar and Sannasiraj [23]. The above examples prove the effectiveness of groin fields as a coastal protection measure.

Offshore Detached Breakwaters

As the name suggests these structures are non-shore connected and lies in the offshore parallel to the coast, their function is similar to that of the arms of the extended breakwaters. The incident wave energy is diffracted by these structures and result in the formation of a calm lee side which are zones of lesser energy dissipation, thereby facilitating the formation of coastal features such as salient and tombolo. The details of the design and specifics are discussed by Sundar and Sannasiraj [24]. A case study is presented by Fujiwara et al. [4] where a series of 12 detached breakwaters protect the Kaike coast in Japan against excessive erosion as shown in Fig. 6.

Fig. 6 Aerial view of beach formation (2003) along Kaike coast [4]

Although the effectiveness of offshore detached breakwaters to combat coastal erosion is proven through various literature, this measure has not been popular in the Indian waters, since it is both cost and labour-intensive measure, which requires heavy machinery, skilled labours, barges, etc. for successful commissioning. It is recommended that coupling detached breakwaters with wave energy conversion devices, such as an oscillating water column [11], which would make the project undertaking more economically viable. Integrating an Oscillating water column (OWC) in a detached breakwater as a coastal protection measure, a study initiated by Ashlin et al. [1] has claimed with thrice the width of the OWC spacing within one of the detached breakwaters exhibited a better performance to an extent of extracting about 2.2 times of the incident wave power. This could be an option in particular for the Islands.

Seawalls

A seawall is usually constructed along the coast and its purpose is to protect the coast by mitigating erosion and flooding. They are usually constructed when the original beach width of the shore is less and if the seaward bed slope is very steep. The seawall structure does not alter the local morpho-dynamics whereas it stabilizes the existing condition by preventing additional damage to the coasts. These structures can be composed of rubble mounds, concrete units, gabions, or plain concrete walls. They can also be of vertical, sloped, or curved surfaces.

It is evident that the conventional sloping face seawalls composed of rubble mounds units require repetitive repair and rehabilitation works at constant intervals as reported by Sundar and Murali [20].

The Shankumukam coast (8° 28′ 49.2″; 76° 54′ 37.7″) in Thiruvananthapuram district is a densely populated region and its shoreline extends to about 4 km, up to the mouth of Veli estuary which is a virgin beach. The rate of erosion is found to be severe during the months of June to August (South-West Monsoon season) as shown in Fig. 7. To combat the seasonal erosion problem construction of a seawall was proposed by the Kerala Irrigation Department as shown in Fig. 8a, with a wide base of 21 m. Sundar and Murali [20] proposed an alternative solution as shown in Fig. 8b with a base width of only 8 m, which makes some part of the beach available during non-monsoon seasons and during peak monsoon season and the designed seawall would perform satisfactorily. The seawall cross section construction along a 6 km stretch of the coast north of Chennai harbour (see Fig. 8c) as discussed by Sundar et al. [18] has withstood the forces due to the great Indian Ocean tsunami of 2004. A comprehensive assessment of the coastal protection measures implemented as a response to the tsunami of 2004 is reported by Sundar et al. [22].

Fig. 7 A view of Shankumukam coast during south-west monsoon

4.3 Soft Engineering Measures

Artificial Reefs

Reefs are natural coral formations on the seabed surface which would actively dissipate incident wave energy approaching the coast by facilitating gradual dissipation and even pre-mature wave breaking. The coral formation promotes the coastal ecosystem by supporting algae and vegetal growth over its surface in addition to supporting fish culture. Since it is difficult to plant corals in desired locations, coral like artificial concrete units is used to serve a similar purpose across desired locations. These concrete units can be precast or cast in-situ and positioned offshore over the seabed at desired lengths and intervals to protect the proximate shoreline.

Vaan island, Tamilnadu: Artificial concrete units (in cross section and plan) as shown in Fig. 9 were used to protect the coast of Vaan island (8° 50′ 17″ N and 78° 12′ 37″ E), in the Tuticorin district off south-east Indian coast. In a period of over four decades the land area of the island was reduced by approximately 85% [5]. From field investigations, Patterson et al. [10] reported that these reefs within six months of installation exhibited remarkable recovery of beach width along the northern side of the island.

Geo-synthetic Seawall

In line with the newly adopted materials and techniques in coastal protection, an innovation cross-section of a seawall comprising different geo-synthetic units such

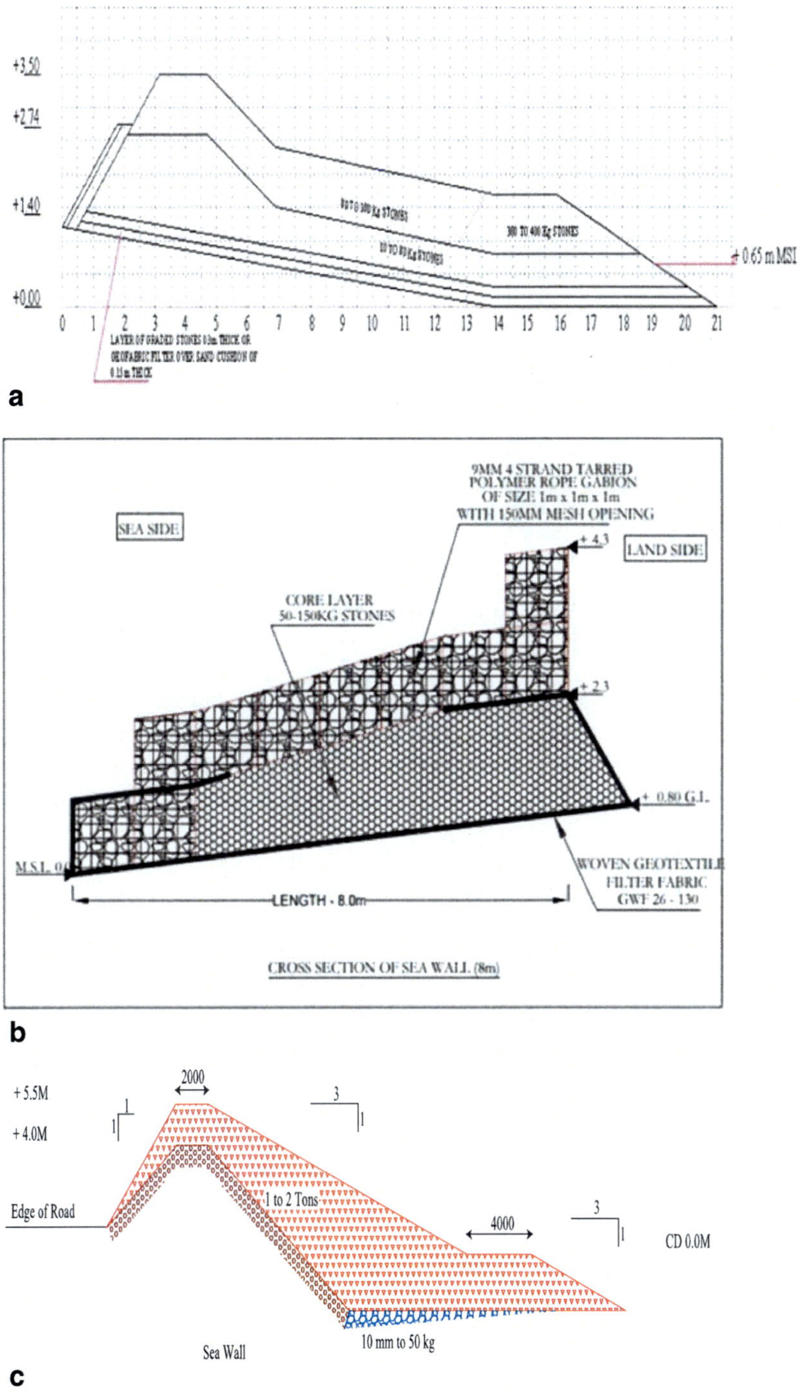

Fig. 8 **a** Seawall cross section at Shankumukam (Proposal 1). **b** Seawall cross section at Shankumukam (Proposal 2). **c** Seawall cross section at North of Chennai harbour [18]

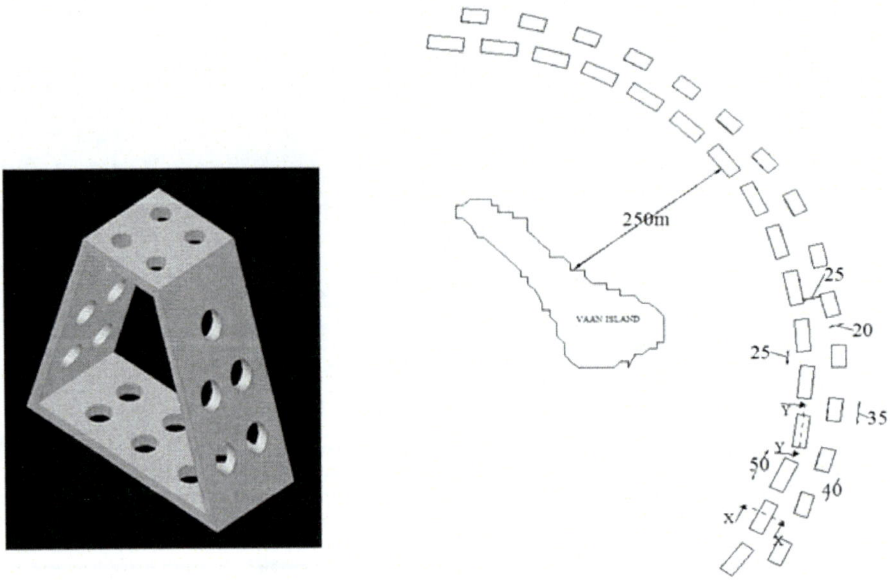

Fig. 9 Cross section and plan of artificial reefs in Vaan island

as geo-synthetic reinforced soil retaining wall (GRW/ barrier force system), geo-bags (small bags and mega bags), geo-cells, geo-grids, and geo-textiles (woven and non-woven) was proposed as shown in Fig. 10. This serves the purpose similar to that of a conventional seawall, without employing rubble stones or concrete units. The

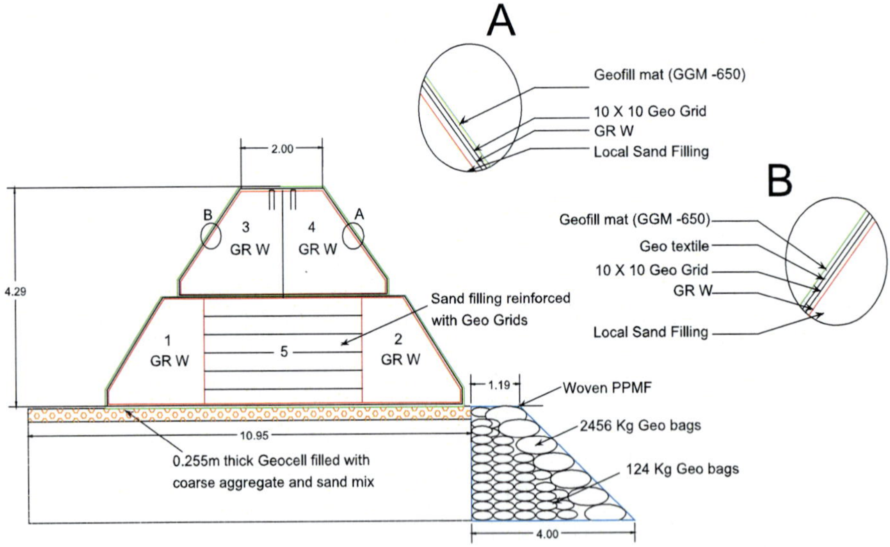

Fig. 10 Geo-synthetic seawall cross section

geo-synthetic seawall was tested for hydro-dynamic stability through physical model studies and geotechnical stability through an analytical model study by Sukanya et al. [17]. Later, this section was implemented along the Pallana beach (19° 17′ 55.19″ N and 76° 23′ 18.55″ E), located in Allepey district of Kerala along the South-west Indian coastline, as a pilot plant extending to about 100 ft in length parallel to the coast. The structure performs satisfactorily well and prevents coastal flooding during monsoon season. The image of the seawall in the field soon after construction in 2015 and later in 2019 is projected in Fig. 11.

Application of Geo-synthetic products like geo bags, geo-containers and geo-tubes to perennial erosion (Fig. 12a) along the Digha beach, Shankarpur, West Bengal was designed and implemented along the coast, the cross section of which is shown in Fig. 12b. The details are reported by Sundar et al. [21]. The protection measure was quite effective as can be seen in Fig. 12c. Unfortunately, due human interference and damages to the geo-tubes lead to its failure around 2011, thus questioning its sustainability.

Fig. 11 A view of geo-synthetic seawall along Pallana, Kerala

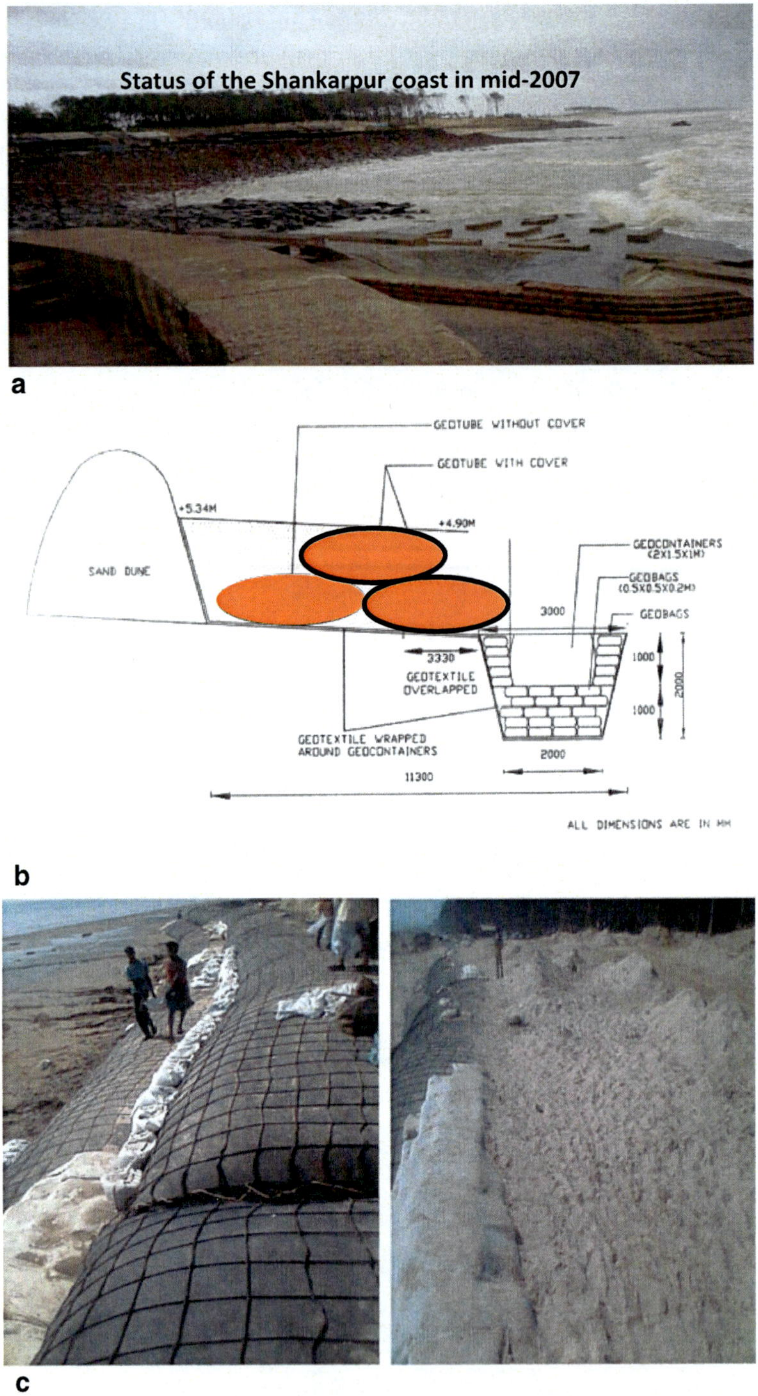

Fig. 12 **a** Status of the Digha coast in 2007. **b** Cross section of protection for Digha. **c** The effectiveness of the Seawall in Digha after construction

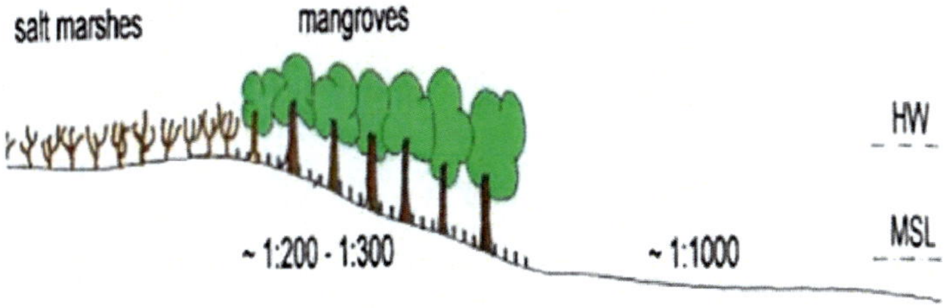

Fig. 13 Vegetation roots prevent scour [16]

Artificial Beach Nourishment

The process of transferring the sediments from the updrift side to the downdrift side by means of slurry pumps at regularly defined intervals to stabilize the beach is known as artificial beach nourishment. The initial cost for procuring and installing high-capacity slurry pumps are high. The allied operation, maintenance and repair costs for such facility is also expensive. This approach is suitable for locations where the updrift accreted coast and downdrift eroded coast are in close proximity, making it easier for the transfer of sediments. A pilot study on the sand by-passing along the Puducherry coast as discussed by Neelamani and Sundaravadivelu [7] was an attempt that initially yielded fruitful results but could not sustain due to several reasons, the main issue being related to the pump capacity and problems in handling.

Vegetation Cover

It is a common observation that the regions of dense mangrove forests along the coast experience minimum or no erosion even during extreme coastal hazards. The root system of the mangroves holds back the sediments and prevents it from getting washed away due to long-shore and/or cross-shore currents, seawater inundation and run-down. Coastal vegetation also influences the resistance against sliding. The roots clearly armour the soil, see Fig. 13 from Schiereck [16]. Noarayan et al. [9] conducted a comprehensive laboratory study on the hydraulic resistance characteristics of a flexible plantation in a laboratory. The view of the laboratory set-up for varying vegetation thickness and spacing is seen in Fig. 14. A relative rigidity parameter for the vegetal growth is proposed in correlation with the friction factor. It was concluded that for ideal flow conditions the incident wave energy can be reduced by up to 90% by means of vegetation cover.

5 Conclusions

A comprehensive review of the hard and soft measures implemented as coast protection measure along the Indian coast has been discussed in this paper. The salient conclusions drawn are

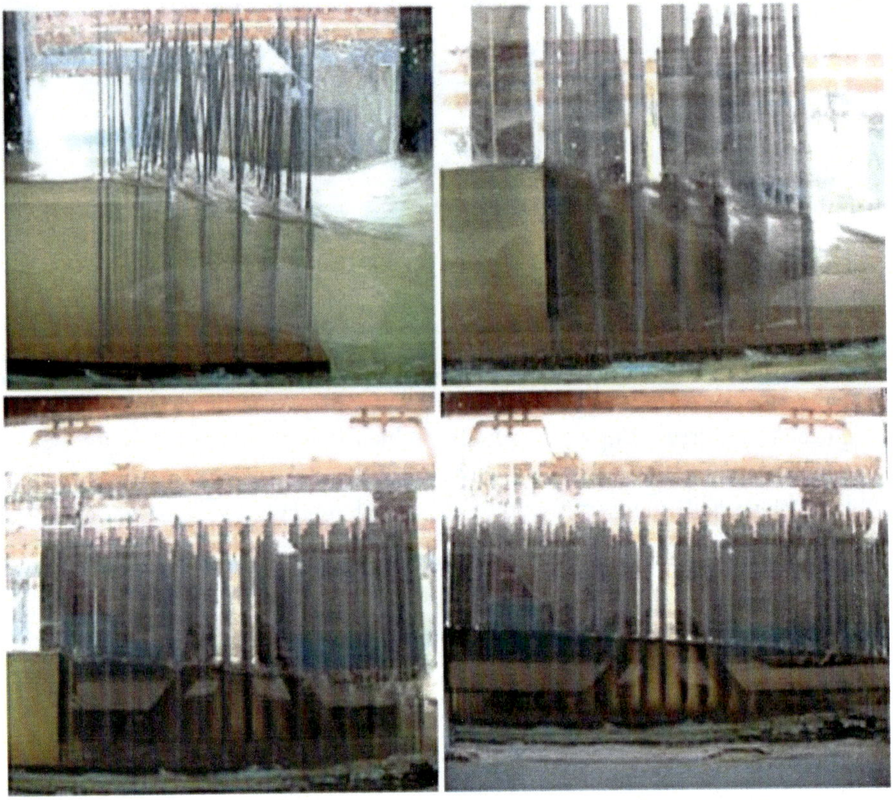

Fig. 14 Typical view of the varying thickness of vegetation tested in the laboratory [9]

The **hard measures**, viz., groin fields along North Chennai harbour have solved a five-decade-old problem. About 30 stretches along the Kerala coast proved to be effective over the past decade has proved its efficiency in protecting the eroding coast. The re-designed Seawalls have proved to be effective. Both the above protection measures have also withstood the forces induced by the tsunami of 2004. Detached offshore breakwaters integrated with wave energy convertor like the Oscillating water column device could be a viable solution for Island an aspect that needs detailed investigations. The depletion of natural rocks leads to dependency on concrete blocks necessitating the exploration of new types of efficient and cost-effective concrete blocks like KOLOS.

The **soft measures** adopted have exhibited mixed responses. Artificial beach nourishment poses problems related to pump capacity, its maintenance and availability of continuous supply of sand. The plantation although eco-friendly can be adopted only as a long-term protection measure. The application of geo-synthetic products adopted has raised concerns on its sustainability particularly the emerging type exposed to sunlight as UV rays poses a challenge. The other problem is human vandalism. Execution and Maintenance of such products need further investigation as skilled labour/simple technology is needed for filling geo-tubes.

Acknowledgements The author expresses his sincere to Public works department of Tamilnadu, Harbour Engineering and Irrigation departments of the Government of Kerala and West Bengal Fisheries Corporation for entrusting the responsibility of providing the coastal protection measures discussed in this paper.

References

1. Ashlin SJ, Sannasiraj SA, Sundar V (2018) Performance of an array of oscillating water column devices integrated with an offshore detached breakwater. Ocean Eng 163:518–532. https://doi.org/10.1016/j.oceaneng.2018.05.043
2. Chandramohan PV, Sundar V, Sannasiraj SA, Arunjith A (2012) Development of 'kolos' armour block and its hydrodynamic performance. In: Proceedings PIANC-COPEDEC VIII, eighth international conference on coastal and port engineering in developing countries, IIT Madras, Chennai, India, Feb 20–24, 2012, pp 1344–1354
3. Coastal Engineering Manual (CEM) (2006) U.S. Army Corps of Engineers, manual, 110-2-1100
4. Fujiwara H, Uda T, Onishi T, Miyahara S, Serizawa M (2012) Prediction of beach changes around artificial reef using BG model. In: Proceedings of 33rd conference on coastal engineering, Santander, Spain, pp 1–12
5. Jayaprakash M, Sivakumar K, Muthusamy S, Krishnamurthy RR, Patterson EJK (2016) Shrinking of Vann Island, Gulf of Mannar, SE coast of India: assessing the impacts. Nat Hazards 84:1529–1538
6. Koerner RM (2012) Designing with geosynthetics, 6th edn. Xlibris Publishing Co., New York
7. Neelamani S, Sundaravadivelu R (2006) Erosion protection of Pondicherry Coast, South East India. J Coast Res 880–883
8. Nicholls RJ, Hanson S, Herweijer C, Patmore N, Hallegatte S, Corfee-Morlot J, Chateau J, Muir-Wood R (2007) Ranking of the world's cities most exposed to coastal flooding today and in the future. Extract from full report. OECD environment working paper no. 1 (ENV/WKP (2007)1)
9. Noarayanan L, Murali K, Sundar V (2012) Role of vegetation on beach run-up due to regular and cnoidal waves. J Coast Res 28:123–130
10. Patterson EJK, Sannasiraj SA, Malleshappa H, Sundar V, Deepak Bilgi S (2016) Integrated approach to protect and restore sinking Vaan Island in Gulf of Mannar, India. In: Proceedings of Arabian coast (3rd international conference on coastal zone engineering and management in the Middle East), UAE, 20–23 Nov, pp 81–83
11. Pilarczyk KW, Zeidler RB (1996) Offshore breakwaters and shore evolution control. A.A. Balkema Publishers, Rotterdam
12. Pilarczyk KW (2000) Geosynthetics and geosystems in hydraulic and coastal engineering. A.A. Balkema Publishers, Brookfield, USA
13. Recio J, Oumeraci H (2008) Hydraulic permeability of structures made of geotextile sand containers—laboratory tests and conceptual model. Geotext Geomembr 26:473–487
14. Recio J, Oumeraci H (2009) Process based stability formulae for coastal structures made of geotextile sand containers. Coast Eng 56:632–658
15. Sanjeevi S, Jayaprakash N (2013) Remote sensing and field studies to evaluate the performance of groins in protecting an eroding stretch of the coastal city of Chennai. In: Proceedings of GIS Ostrava, geoinformatics for city transformation
16. Schiereck GJ (2001) Introduction to bed, bank and shore protection. Delft University press, Postbus 98, 2600 MG Delft. ISBN 90-47-1683-8
17. Sukanya R, Sundar V, Sannasiraj SA (2016) Stability and settlement analysis of compound geo-synthetic materials in coastal protection measure. In: Proceedings of ninth international conference on coastal and port engineering in developing countries (PIANC-COPEDEC IX)

18. Sundar V, Sannasiraj SA, Nasar T, Balaji R (2004) Anti-sea erosion work at coastline north of Royapuram fishing harbour. In: International seminar on coastal area construction management, Mumbai, pp 57–61
19. Sundar V (2005) Planning of coastal protection measures considering the effects of tsunami along the coast of Tamil Nadu, India. In: Proceedings of national workshop on tsunami effects and mitigation, pp 211–230. ISBN: 81-7764-93-96
20. Sundar V, Murali K (2007) Planning of coastal protection measures along Kerala coast. Report submitted to the State Government of Kerala by IIT Madras
21. Sundar V, Maiti DK, Sannasiraj SA, Venkatraman M (2009) Geosynthetic application for coastal protection at Shankarpur, West Bengal, India. In: Proceedings of 5th international conference on Asian and Pacific Coasts (APAC 2009), Singapore, pp 58–64
22. Sundar V, Sannasiraj SA, Anil Kumar G (2014) Post-facto evaluation coastal protection measures along tsunami hit Kerala coast of India. In: Proceedings of Indo-Japan workshop on river mouths, tidal flats and lagoons, IIT Madras, Chennai, India, pp 1–20
23. Sundar V, Sannasiraj SA (2018) Coastal protection along vulnerable stretches along the south Indian peninsula (Chapter 46). In: Kim YC (ed) Handbook of coastal and ocean engineering, vol 2, 2nd edn. World Scientific, Singapore. http://doi.org/10.1142/10353
24. Sundar V, Sannasiraj SA (2019) Coastal engineering: theory and practice (Chapter 4). In: Advanced series on ocean engineering, vol 47. World Scientific. http://doi.org/10.1142/11148

Issues Related to Erosion and Its Effects on Structures in Visakhapatnam

Chirla N. V. Satyanarayana Reddy

Abstract Soil erosion caused by surface runoff poses serious problems to the stability of structures built in sloping ground. Erosion of soil from hill slopes result in soil slides with boulders/disintegration of soil mass and affects the foundations laid in the vicinity of hill slopes. Deep excavations in clayey silty sands for basement floor constructions are challenging as they appear to stand vertical in dry state and any saturation of soil due to rainfall or subsurface seepage and associated erosion result in sudden sliding of soil mass from the excavated surface. Also, the Coastal cities are getting vulnerable to erosion of beaches caused by sea erosion due to development of ports and improper nourishment activities undertaken by the port and local authorities. In the present paper, issues related to beach erosion, effect of soil erosion from surface runoff in sloping ground and deep excavation failures pertaining to Visakhapatnam are presented. The various remedial measures to check the erosion are discussed.

Keywords Soil erosion · Beach erosion · Stability of soil · Hill slopes · Excavations

1 Introduction

Erosion of soil is a phenomenon of disintegration of soil mass caused by surface/subsurface water and washing away of soil fines/soluble material with the seeping water. Erosion of soils mainly occurs in fine-grained cohesionless soils and granular material bonded with weakly cemented water soluble binding material. A soil's inherent erodibility is a major factor in erosion prediction and land-use planning and dependents on its infiltration capacity and capacity to resist detachment and transport by rainfall and runoff [1]. Erosion of low plastic fines of gravelly soil in slope leads to instability of slope due to sliding of disintegrated mass. Usage of

C. N. V. Satyanarayana Reddy (✉)
Department of Civil Engineering, College of Engineering, Andhra University, Visakhapatnam, India
e-mail: prof.cnvsreddy@andhrauniversity.edu.in

C. N. V. S. Reddy and S. Sassa (eds.), *Scour and Erosion Related Issues*, Lecture Notes in Civil Engineering 177,
https://doi.org/10.1007/978-981-16-4783-3_5

industrial wastes as construction materials is also posing threat to constructed structures due to their high susceptibility to erosion. The embankments/reinforced earth retaining walls were built using fly ash without adopting proper measures to check its erosion, as construction materials are leading to instability due to erosion of material due to subsurface flow.

Soil erosion issues are common in coastal cities with undulating topography. The presence of silty sand/fine sand aggravates the problem of erosion. Hill slope constructions seriously suffer from erosion of soil from slopes if proper measures are not adopted. Erosion of soil in hill slopes destabilizes the structures over a period of time and lead to the collapse of structures. Wischmeier and Mannering [1] reported that the most erodible soils are those rich in loam and fine sand. As long as the flow velocity is slow (less than 0.25 m/s), the erosion problem does not arise. Hence, measures are to be taken to slow down the surface flow to prevent erosion. Erosion on hill slopes shall be controlled using natural vegetation, turfing, vegetation mats, stone pitching, geonets, gabion mattresses and retarding flow by breaking the continuity of slope through the introduction of berms with retaining walls at every 3–4 m interval.

Deep excavations made in the construction of basement floors of high rise structures in urban areas are leading to instability of neighbouring structures as proper soil retention measures are not being adopted. Such failures are happening in low plastic granular soil which appears to stand vertical in deep excavations and collapse when they get saturated due to rain or subsurface water seepage. Hence, it is essential to highlight the importance of supporting systems [2, 3] for avoiding soil slides from excavated faces and to ensure the safety of working personnel and stability of neighbouring structures. Further, in coastal cities with ports, erosion of beaches by sea waves is caused by the action of sea waves. Deep dredging at ports to facilitate movement of large vessels is the major cause for erosion in adjacent the region apart from erosion caused by aggressive waves hitting the coast. Also, any constructions made extending into the sea lead to erosion issues.

In the present paper, the effects of soil erosion on stability of structures in Visakhapatnam, a city on the east coast of India are presented.

2 Effect of Erosion of Hill Slope Material on Foundation Stability

Due to scarcity of plain ground availability in Himalaya mountain ranges, eastern and Western Ghats regions, hill slope constructions are common in India. Constructions are increasing at a faster rate in these areas and posing problems to engineers in terms of their stability. Erosion of soil from heavy rains abutting the foundation soil of structures in sloping ground causes failure of structures. Soil erodibility is governed by clay content, particle size distribution, compaction, permeability, shear strength parameters, etc. of the soil. Due to unprotected soil slope and greedy people excavating the neighbouring site soil in hill slope for their need steepen and subsequently erode during rains and lead to failure of foundations of structures [4]. Visakhapatnam city with several constructions on hill slopes is experiencing foundation failures due to erosion of soils. one such failure is presented below wherein RC cantilever retaining wall of an apartment in VUDA Colony, Marripalem, Visakhapatnam constructed in a hill slope collapsed due to improper soil support. 6 m high rear side RC retaining wall of an apartment built in hill slope failed by overturning due to bearing failure (Fig. 1). The retaining wall had backfill of about 5.4 m and was supporting the parking area of stilt floor. The retaining wall was founded at 1.5 m depth below ground surface in clayey gravel with properties listed in Table 1.

The failure resulted from excavation of foundation soil of retaining wall by local people and subsequent erosion of soil during heavy rains. Prior to failure, the foundation soil on the toe side of retaining wall was standing almost vertical for 2 years and no foundation soil stabilizing/retention measures are taken by the residents and led to erosion and disintegration of clayey gravel to cause bearing failure in the toe region. IS 1904-1986 recommends a clear cover of 90 cm beyond footing edge to sloping surface and further, the load dispersion line from edge of footing shall not cross the hill slope surface. Critical examination of these regulations reveals that they cannot ensure availability of soil required for mobilization of bearing capacity

a) View of failed retaining wall

b) Damaged apartment in construction due to retaining wall collapse

Fig. 1 **a** View of failed retaining wall. **b** Damaged apartment in construction due to retaining wall collapse

Table 1 Engineering properties of foundation soil

S. No.	Engineering property	Value
1	Specific gravity	2.68
2	Particle size distribution	
	(i) Gravel (%)	66
	(ii) Sand (%)	21
	(iii) Fines (%)	13
3	Plasticity characteristics	
	(i) Liquid limit (%)	30
	(ii) Plasticity index (%)	11
4	IS Classification (as per IS 1498-1978)	GC
5	In-situ density (g/cc)	2.14
6	Natural moisture content (%)	5.8
7	Shear parameters	
	(i) Cohesion (t/m^2)	1.2
	(ii) Angle of internal friction	32°

as Zone 3 (Zone of mixed shear) extends to a distance of 3–4 times size of footing in granular soils. Lack of sufficient soil cover beyond the failure surface/shear zones results in non uniform mobilization of bearing capacity with lesser value below the toe which experiences higher base pressure from footing of retaining wall.

The effect of loss of soil above the base level of foundation will reduce bearing capacity due to lack of surcharge pressure '$\Upsilon.D_f$'. Further, loss of soil to the right of a vertical plane passing through edge of footing results in reduced bearing capacity contribution from zones of radial and linear shear in toe side footing portion. The loss of soil beyond the toe of retaining walls results in reduced bearing. The retaining wall base slab exerts more base pressure at the toe than the heel. Since erosion of soil near the toe of retaining wall results in reduced bearing capacity, gradually the foundation experiences tilting and finally lead to collapse of the wall. Such failure is evident from the analysis of bearing capacity presented below.

In the present study, the effect of soil loss on bearing capacity is assessed with respect to the foundation of the affected cantilever retaining wall on hill slope comprising of Clayey Gravel (properties given in Table 1). The Section of retaining wall is shown in Fig. 2.

The reduction in bearing capacity on the toe side is estimated by plotting the shear zones as per Vesic's theory in AUTOCAD for the retaining wall section shown in Fig. 2. The percentage reduction in areas of shear zones II and III is used to make deduction for passive resistance of soil in 3rd term of bearing capacity.

The safe bearing capacity of soil is determined under normal conditions (prior to loss of soil cover beyond the toe of footing) and with loss of soil above base of footing and beyond the vertical plane through the toe of the wall. The analysis yielded safe bearing capacities of 19.2 and 14.9 t/m^2 at the toe (with loss of soil cover) and

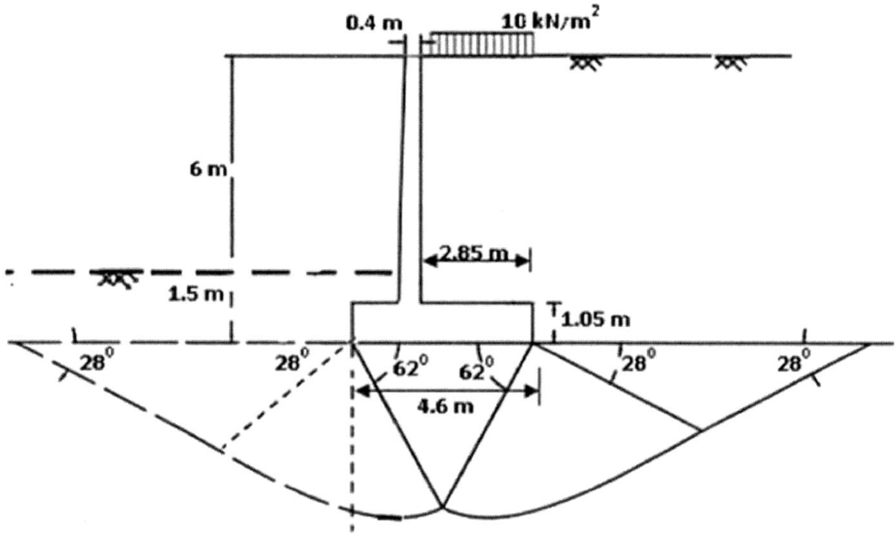

Fig. 2 Section of retaining wall with shear zones

59.2 and 34.1 t/m^2 at heel in dry and saturated conditions respectively. Thus there is significant reduction in bearing capacity of soil from the heel to the toe side. The maximum base pressure at the toe of the wall is determined as 20.8 kN/m^2 at the toe from stability analysis considering a surcharge pressure of 10 kPa. The analysis revealed that the base pressure exceeded safe bearing capacity of eroded foundation soil at the toe of the wall and led to soil bearing failure and resulted in failure of retaining wall by overturning.

Hence, efforts are to be made to protect the foundation soil beyond the toe of the footing in hill slope for stability of retaining wall. The problem of instability further aggravates under seismic/wind loads. There is a need for revising the code stipulations for maintaining sufficient soil cover beyond the edges of footings in slopes as the present guidelines cannot ensure stability of footings of retaining walls of moderate to large widths.

3 Erosion Related Issues in Excavated/Natural Hill Slopes

High rise constructions with cellar and sub cellar floors in already built-up areas are increasing in Visakhapatnam. The excavations for creation of basement floors without adoption of proper soil retention systems/methods are not only resulting in instability of soil from excavated faces and seriously affecting the stability of neighbouring structures, but also affecting the lives of working personnel under the large sliding soil mass.

Fig. 3 Deep excavation failure in clayey silty sand at a construction site in Visakhapatnam

Four migrant workers were buried alive at a construction site near the upscale VIP Road in Visakhapatnam on September 18, 2014, after the soil caved in at a construction site (Fig. 3). The site had vertical excavation of nearly 6 m in clayey silty sand (properties given in Table 2) for the basement floors construction. Large mass of soil from excavated face slid down suddenly when further excavation was done for basement wall foundations and some of the workers got buried in foundation trenches. There was overnight rain and soil was in saturated state at the affected site.

For the Clayey silty sand at the site, in dry state, the critical depth of vertical excavation is determined as 7.6 m from Taylor's stability chart [5] and hence, no instability occurred when excavation depth reached 7.5 m. However, in saturated state, due to significantly reduced cohesion, stable excavation depth is 3 m only as per Taylor's stability number. Hence, soil from top 4.5 m depth caved into excavation and led to the mishap. Hence, vertical excavations without proper retaining system shall not be attempted in such soils as they mislead the construction engineers with their characteristic dry strength induced by dried binding plastic fines. The bonding of particles greatly decreases upon saturation due to reduced cohesion of plastic fines and erosion of binder material from excavated faces. Excavations in such soils shall be done by adopting nailing system [6, 7] with nail heads embedded in shotcrete slab to check erosion of soil from excavated soil mass.

Adequate surface drainage is also necessary in new excavations as well as in maintenance of constructed slopes where movement has already occurred. The design of cut slopes should always take into consideration the natural drainage patterns of the area and the effect that the constructed slope will have on these drainage patterns. Erosion and disintegration of soil mass due to saturation are responsible for caving in of soil from excavated slopes in low plastic sandy or gravely soils. Control of water flow on excavated soil surfaces and prevention of water seepage into soil from head of slope help in avoiding sliding actuating forces greatly and ensure the stability of unsupported soil slopes.

Table 2 Engineering properties of soil at affected area

S. No.	Property	Value
1	Specific gravity	2.66
2	Grain size analysis	
	(a) Gravel (%)	0
	(b) Sand (%)	83
	(c) Fines (%)	17
3	Plasticity characteristics	
	(a) Liquid limit (%)	23
	(b) Plastic limit (%)	18
	(c) Plasticity index (%)	05
4	IS Classification (as per IS 1498-1978)	SC-SM
5	(a) In-situ dry density (g/cc)	1.77
	(b) Natural water content (%)	1.8
	(c) Saturated density	2.10
6	Shear parameters I. Natural state	
	(a) Cohesion (kN/m^2)	21
	(b) Angle of internal friction	30
	II. Saturated State	
	(a) Cohesion (kN/m^2)	12
	(b) Angle of internal friction	28
7	Stable height of vertical excavation (m)	
	(a) Dry state	7.8
	(b) Saturated state	3.4

4 Beach Erosion

Surging of sea occurs due to several factors such as deep dredging activities at ports, construction of harbours in near shore, construction of sediment-trapping upland banking, construction of groins and jetties, etc. Port authorities generally administer the beach nourishment through pumping of dredged sand towards the beach to check the erosion at vulnerable beaches in the neighbourhood. The nourishment of beach is to be carried out on regular basis and any deviation results in eroded shores.

Visakhapatnam port is regularly enhancing its activity by taking up deep dredging in harbour and construction of new jetties and also new port has been set up at Gangavaram. Subsequently, during January 2014, a sea surge took place in Rama Krishna Beach by about 30–40 m over a length of 400 m and the sea waves hit against the soil fill supporting footpath and neighbouring pavement (Fig. 4). The subsoil supporting the footpath and adjacent pavement is comprised of fine sand (non-plastic material). The waves from the sea surge have resulted in base erosion

Fig. 4 Eroded beach affecting neighbouring footpath at Visakhapatnam

and migration of moisture into soil due to capillarity. As a result, the fine sand mass collapsed and led to damage to footpath.

As an immediate measure, for preventing further damage to footpath and pavement adjacent to beach, the authorities have dumped Coarse gravel and bouldery material (which do not disintegrate in presence of water) to check further erosion and stabilize the existing fine sand. Local authorities have consulted experts from NIOT, CWPRS, Pune and abroad to get permanent/ long term solutions to mitigate the erosion problem and various causes for erosion and mitigation methods have been floated. The eroded beach is recovered after extensive beach nourishment done before the commencement of International Fleet Review (IFR) 2016, an international maritime exercise hosted by Indian Navy. Since then, the severity of shore erosion problem is not observed much as beach nourishment is being done on regular basis.

However, shore protection measures are required to protect the subsoil adjacent to beach supporting various infrastructures in the neighbourhood against erosion from the approaching waves in the event of sea surge. Among several options for prevention of shore erosion, the following systems are generally adopted in different countries.

- Filter Type Structures
- Gabion Structures
- Geo tube Walls
- Groins.

Filter type structure (Fig. 5) reduce the energy of the incoming waves and retain the soil in place. Reduction of the energy of waves is accomplished by the flat slope (2H:1V) and the rough surface. Filter layer on sloped bund comprises of 150–200 mm thick layer stones of varying sizes laid over a geotextile fabric placed on the slope. Over the filter layer, armour stone layer is provided with larger sized rough angular rock material. The provision of armour stones is required only up to a height of 1.5 m above the toe of revetment as height of wave approaching is 0.2–0.3 m at the

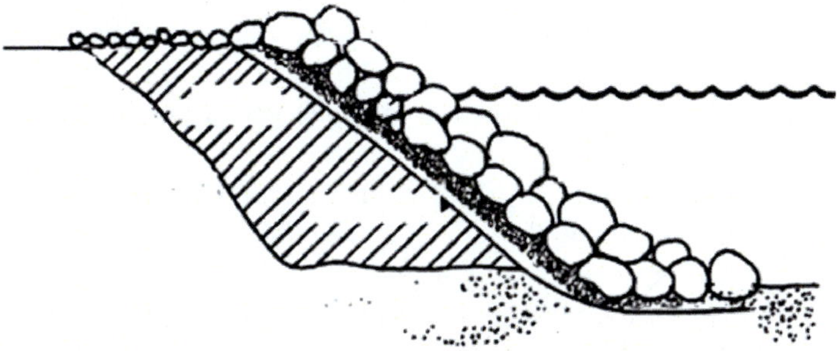

Fig. 5 Stone revetment system for shore erosion protection

affected beach in Visakhapatnam. Toe protection helps in preventing settlement and protecting the edge of the revetment from washing away. An overtopping apron of 250–300 mm thickness is generally provided over 2–3 m length on top of revetment for protection during overtopping.

Gabions are rectangular wire baskets filled with durable stone material. The baskets are made of galvanized and polyvinyl chloride (PVC) coated steel wire in a hexagonal mesh. 100–150 mm size crushed rock/stones are used to fill the baskets. Figure 6 shows gabion structure for erosion control of beach. There are two types of gabions namely upper level Baskets and mattresses. Mattresses are rock-filled baskets of 225–300 mm thickness and provide proper foundation for the upper level baskets. Upper level baskets have lengths of 150 mm, 225 mm and 300 mm lengths and heights of 300 mm, 450 mm and 900 mm. At the construction site, gabion baskets are unfolded and assembled by lacing the basket edges with wire. Individual baskets are then laced together, stretched and filled with stones. The lids are closed and wired to other baskets to form large heavy mass that is not as easily moved by waves like single stones.

The construction of gabions is accomplished without heavy equipment. The structure is flexible and continues to function properly even if the foundation settles. Adding stones to the baskets is an easy maintenance procedure. The cost of gabion

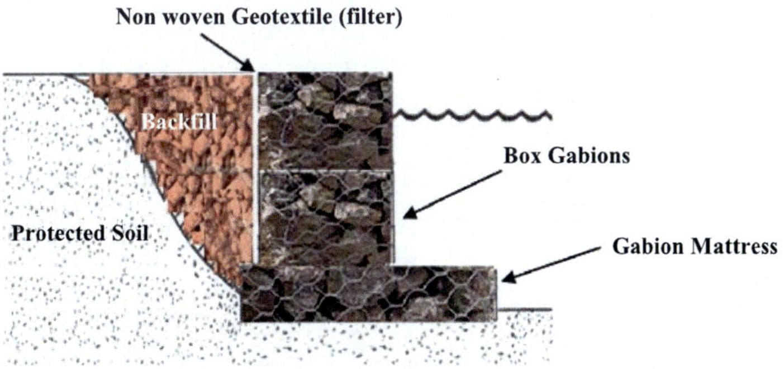

Fig. 6 Wall type erosion control system with gabions

system is low compared to other protection methods if rock material is locally available. Gabion baskets may open under heavy wave action releasing the stones and scattering them. Also, waterborne debris, cobbles, ice and foot traffic can damage the baskets. Corrosion of baskets placed in saltwater gets initiated even with the small defect in the protective coatings.

Geotubes are tubes (Fig. 7) made of synthetic geotextiles and are filled with sand extracted from sea bed if filling granular material not dissolving in water is not locally available. They are 3–5 m in diameter and available in different lengths. The granular soil filled geotubes serves as massive retaining systems. The geotubes are relatively flexible compared to conventional retaining walls and absorb the energy of approaching waves. The geotube wall is shown in Fig. 4. Placement of geotextile tubes system in sandy beaches prevents coastline erosion and nourishes the beach [8]. Installation of geotube system/geogrid mattress at two or three locations parallel to the coast aids in regaining the part of the eroded beach over a period of time.

Groins are the rubble stone masonry walls built into the sea for control of erosion and for creation/regain of the eroded beach. But, the gain in eroded beaches always results in erosion of down drift side beaches [9]. Also, groins alter the natural state of beach. Hence, usage of groins is banned in some countries.

5 Conclusions

Based on the analysis of case studies presented in paper, the following conclusions are drawn.

1. Deep excavations shall not be made in clayey silty sand without adopting proper soil retention systems as soil is prone to sudden collapse upon saturation under rain/subsurface seepage water though they appear to stand vertical to larger depths in excavation. Also, it is essential to plan surface drainage to prevent flow of water along the excavation and seepage into soil adjacent to excavation.
2. Adequate soil cover beyond the edge of footing shall be maintained to ensure uniform mobilization of bearing capacity under the footing. Further, erosion control measures are to be adopted to protect the foundation soil in slopes using vegetation mats, stone pitching, etc. Any tilting of foundation in the vicinity of slopes requires immediate attention to avoid failures as it is a sign of loss of proper support/ bearing from soil.
3. Beach erosion control measures are to be adopted if the problem persists despite regular nourishment being done by the authorities. Flexible (soft) systems such as geotube walls, filter type structures with Geosynthetics and gabion mattresses shall be preferred over the rigid groin structures.

Fig.7 Geotube wall for erosion control system for beach

References

1. Wischmeier WH, Mannering JV (1969) Relation of soil properties to its erodibility. Soil Sci Soc Am J 33(1):331–337
2. Holtz RD, Schuster RL (1996) Landslides: investigation and mitigation. Transportation Research Board special report 247, National Academy Press, Washington, D.C., pp 439–473
3. IS 14458 (Part 1)-1998. Retaining wall for hill area—guidelines—Part 1; Selection of type of wall. Bureau of Indian Standards, New Delhi
4. Satyanarayana Reddy CNV (2011) Case studies of recent geotechnical failures in Andhra Pradesh. In: E-Proceedings of 1st young geotechnical engineers conference of A.P, IIIT Hyderabad
5. Taylor DW (1937) Stability of earth slopes. J Boston Soc Civ Eng 24(3):197–246
6. Gassler G, Gudehus G (1981). Soil nailing—some aspects of a new technique. In: Proceedings of 10th international conference soil mechanics and foundation engineering, Stockholm, vol 3, pp 665–670
7. Juran I (1987) Nailed-soil retaining structures: design and practice. In: Proceedings of transportation research record 1119. Transportation Research Board, Washington D.C., USA, pp 139–150
8. Lee SC, Hashim R, Motamedi S, Song K (2014) Utilization of geotextile tube for sandy and muddy coastal management: a review. Sci World J 2014(Article ID 494020). http://doi.org/10.1155/2014/494020
9. Galgano FA, Leatherman SP and Douglas BC (2004) Inlets Dominate U.S. East Coast Shoreline Change. J Coast Res
10. IS 1498-1970. Classification and identification of soils for general engineering purposes. Bureau of Indian Standards, New Delhi
11. IS 1904-1986. Code of practice for design and construction of foundations in soils: general requirements. Bureau of Indian Standards, New Delhi
12. Vesic AS (1973) Bearing capacity of shallow foundations. In: Foundation engineering handbook. Van Nostrand Reinhold Company, New York, pp 121–147

Erosion Control Solutions with Case Studies

Check for updates

Paolo Di Pietro and Ratnakar R. Mahajan

Abstract River engineering requires to start from the location and characteristic of a watercourse like in the mountains, or at the foothills (mid slope), or in the valley, close to the river outlet. Soil erosion in three different regimes varies due to different soils and different morphology of rivers. Soil erosion phenomena follow the principle that instability starts when shear stresses become critical generating soil particles to get to incipient motion with the flow. Gabions and mattresses made of double twisted wire mesh provide shear resistance due to confinement of stones and have been used as erosion protection for many years. Design approach of mattresses is based on a most recent research carried out at Colorado State University in 2019. This paper presents the new design approach along with performance limits of new type of mattress tested, as per ASTM D6460, based on observation of the mattress stability with respect to flow through mattress layer, combining the stability of the stone and its ability to control soil erosion underneath. Also, five case studies are presented covering different solutions which demonstrate performance and durability of mattresses.

Keywords Erosion control · Mattress · Shear stress · Gabion

1 Introduction

River engineering is the process of planned human intervention in the course, characteristics, or flow of a river with the intention of producing some defined benefit. People have intervened in the natural course and behavior of rivers since old times to manage the water resources, to protect against flooding, or to make passage along or across rivers easier. From Roman times, rivers have been used as a source

P. Di Pietro
Officine Maccaferri S.p.A, Via Kennedy 10, 40069 Zola Predosa, BO, Italy
e-mail: p.dipietro@maccaferri.com

R. R. Mahajan (✉)
Maccaferri Env Sol Pvt Ltd, Gurugram, India
e-mail: r.mahajan@maccaferri.com

C. N. V. S. Reddy and S. Sassa (eds.), *Scour and Erosion Related Issues*,
Lecture Notes in Civil Engineering 177,
https://doi.org/10.1007/978-981-16-4783-3_6

of hydropower. In the late twentieth century, river engineering has had environmental concerns broader than immediate human benefit, and some river engineering projects have been concerned exclusively with the restoration or protection of natural characteristics and habitats.

River engineering techniques include various constructional techniques like bank protection, construction of embankments, dams, reservoirs, weirs, etc. These techniques are used to solve site-specific problems. Structures constructed for river engineering can be divided into rigid, semi rigid, and flexible depending on the material used for construction and behavior when subjected to forces. Rigid structures are constructed using RCC or PCC. These structures do not tolerate movements. Flexible structures can adjust settlements without undergoing destructions. For constructing these structures across and along the length of the river different materials are used (steel wire mesh, polymer mesh, and tree trunks). Galvanized steel wire is used to manufacture boulder crates which are hand woven mechanically woven and welded.

The first application of gabions by Maccaferri was built more than 120 years ago, along the river Reno crossing the small town of Casalecchio, in the vicinity of Bologna. This early application was built in 1894 to repair the banks due to a breach caused by a major flood (Fig. 1). The structure is the first example of steel wire mesh used to confine loose stones in order to reduce their required volume. The successful performance of this first riverbank application encouraged Maccaferri to further develop steel wire mesh gabions with basket shape to allow them to connect together (Fig. 2).

Gabion applications start becoming attractive as engineered systems to control erosion, due to their ability to combine strength, flexibility, and a natural pleasing outlook at the same time.

The increasing use of these natural systems leads to an interest to investigate their engineering features through research programs aimed at developing design criteria, to learn their performance as protective systems against hydraulic and geotechnical instability. Among their best features, the environmental friendliness and the ability to integrate with the surrounding environment becomes an attractive subject of research of the 90s.

1.1 Basic Principles of Erosion

Structures in river works are based on a choice of theories combining experience. Also, the know-how gained through various structures and best management practices assuming that the effects of nature (rainfall, flooding, storms, etc.) are described by statistical data. It is required to do a risk assessment understanding the project conditions beyond the data considered/assumed. A modern and correct approach to river engineering requires to recognize two main types of requirements: (i) hydraulic requirements and (ii) environmental requirements.

Hydraulic requirements which includes

Fig. 1 Bank protection along river Reno, Casalecchio

- an assessment of the stability of a solution, both from geotechnical and hydraulic point of view,
- the use of correct and sound engineering practices,
- the use of design methods based on risk assessment,
- consideration on how erosion will affect the surrounding environment. In other words, to use an integrated river management approach.

Environmental requirements consider river as a living environment. These requirements consider:

- the river is a dynamic environment. It is necessary to consider that it will change over time,
- the application of river engineering principles shall minimize changes in the morphology and in the nature of the watercourse,
- to adopt protection measures with proven ability to re-establish the natural conditions. To use "sustainable systems",
- to provide solutions aimed at balancing the eco-systems.

Fig. 2 Evolution of gabion

River engineering requires to start from the location and characteristic of a watercourse like in the mountains, or at the foothills (mid slope), or in the valley, close to the river outlet (Fig. 3). The three regions will be characterized by different climates, different soils, and different morphology.

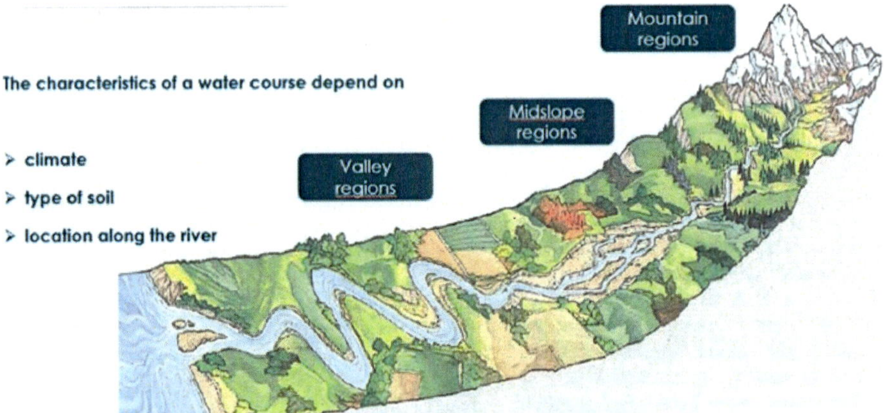

Fig. 3 Regions of river

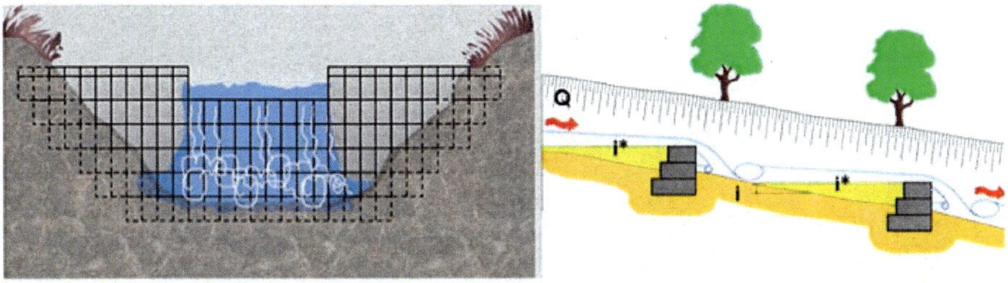

Fig. 4 Weir structure to control gradient of watercourse

River engineering application structures can be divided into three categories: rigid, semi rigid, and flexible structures depending on the behavior. Also, river engineering structures can be classified as: (a) Longitudinal structures and (b) Transverse structures based on location of the structures. Longitudinal structures are provided along the course of the river and transverse structures are provided across the river length or flow.

In mountainous regions, watercourses will be characterized by steep gradients (5–10%), with deep erosion frequently characterized by abundant and coarse bed load. Structures used to prevent erosion in these regions are grade control weirs, sometimes combined with secondary longitudinal bank protections (Fig. 4).

In the mid region of river slope gradients become moderate (1–5%), and river sections start widening, forming a low flow section (a thalweg) and a berm on the lateral sides, which will overflow during the high floods only. Sometimes, during dry weather small islands or rapids in short reaches are formed. Typical structures in this region may be weirs, with a wider crest, combined with bank protections alongside, and small groynes to facilitate sedimentation on curved reaches.

Valley regions are the last segment of the river system, where the watershed has collected most part of the rainfall volume. Rivers are typically characterized by low gradients (less than 1%), low velocities (less than 3 m/s), and large discharges (more than 1000 cum/s). Rivers have a tendency to meander in this region. Interventions in this region will primarily to stabilize the riverbanks (on bends or straight reaches). These interventions favor sedimentation in sections where the meandering tends to progress.

2 Typical Situations and Erosion Control Solution

The problem of floods in natural watercourses requires protection of the riverbank, or to build a new embankment (dyke). Alternatively, where enough area is available, an encroachment area where the water may temporarily flow into a detention basin can be adopted. In mountains option of reducing gradient can be considered. In other cases, a cross section re-profiling may be a reasonable option.

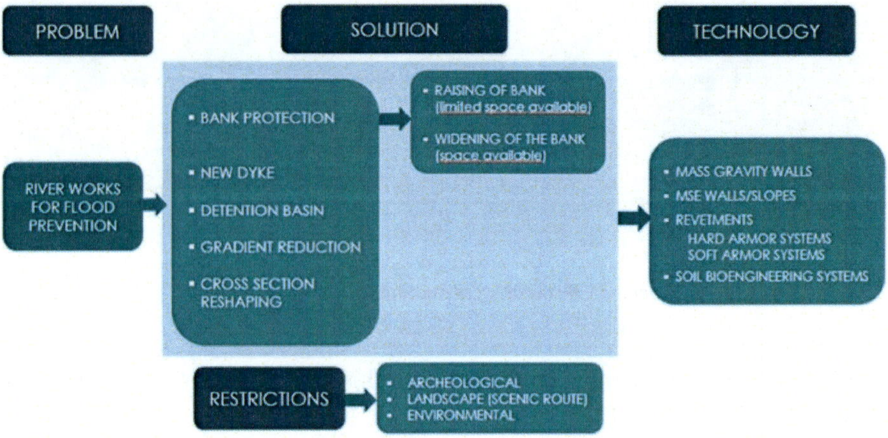

Fig. 5 River works problem and solution

In case of bank protection required, the top of the slope may be sometimes raised (when little space is available on the sides), or otherwise, the section may be widened to convey a larger amount of discharge within the river section. While adopting any of the possibilities there could be a possibility to encounter site limitations, like archeological or landscape restrictions or other environmental issues refer Fig. 5.

Erosion and sedimentation in watercourses are inevitable processes and it is required to understand their nature. Any man-made environment will modify the watercourse which may result in scour. In order to avoid this, erosion control using sustainable systems shall be adopted which will yield and recover the damaged environment and not just fixing the problem.

Type of problem can be started for finding solution including case and effect result of a problem. For example, drainage resulting in surface runoff occurring on natural or artificial slopes leading to erosion due to lack of vegetation or soil instability or seepage, resulting in reduced soil strength, in embankments and slopes. Wet and dry cycles or temperature variations resulting in the formation of cracks in cohesive slopes result in progressive debris.

Ultimately, when it comes to the choice of the most appropriate type of protection, the installation of the protection (whether above or below the waterline) will play an important role in the choice of the solution.

In dry conditions above water level, various options are available from light revetments to hard armor heavy-duty linings like mattresses or gabions. In wet conditions below water level mostly heavy-duty protections are required. Underwater installation require the use of special equipment and sometimes divers are needed, additionally, the inability to connect units with each other will limit the choice of solutions. In most cases, linings with Reno Mattresses or Gabions launched by crane and/or by a barge are adopted. Furthermore, when erosion on slopes is the cause of geotechnical instability, retaining structures shall be used.

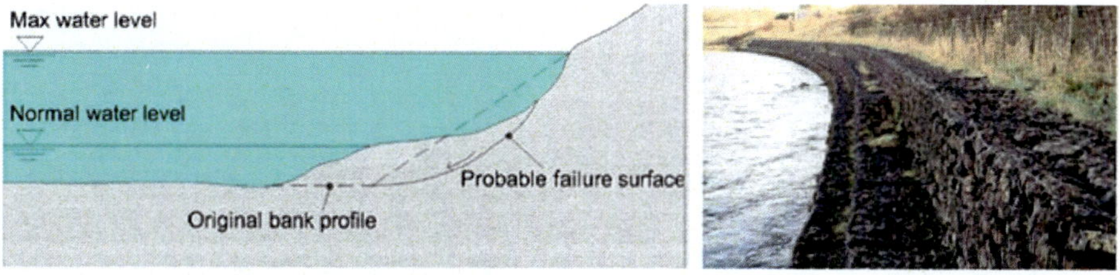

Fig. 6 Instability of riverbank and solution with gabion or mattress

Mass gravity wall of gabion or mechanically stabilized earth walls (MSEW) with launching apron can be used as retaining structures. However, since MSEW requires proper soil compaction, installation demands dry conditions. On the contrary, Gabions and Reno mattresses allow being launched as individual (or multiple) units under the water surface.

2.1 Riverbank Erosion

Figure 6 presents instability generated along the banks due to erosion at the toe and creating a failure surface. Typical solutions for this problem combine gabions or mattresses depending on the angle of the slope and expected shear stress.

2.2 Erosion at Bridge Locations

When infrastructures cross rivers there are requirements of erosion control. Typical situation arises at bridge pier and abutment locations. Pier foundations in highly erodible riverbeds can be subject to instability as piles are progressively exposed to the flow. These situations are still quite critical as rigid supporting elements like piles are in contact with loose erodible soil. Figure 7 present examples of progressive erosion underneath a bridge due to a lack of protection.

Usually, rip rap along abutment may be provided. Gabions and mattresses as flexible erosion protection works manage transition and avoid abrupt changes. Figure 7 present abutment and pier protection using gabion or mattress which require lesser thickness than rip rap due to confinement of stones in wire mesh.

2.3 Guide Bunds

Guide bunds are structures that divert the water to flow in the central zone of the river, to avoid river meandering across the large floodplain with risks for all piers

Fig. 7 Bridge abutment, pier, and floor protection

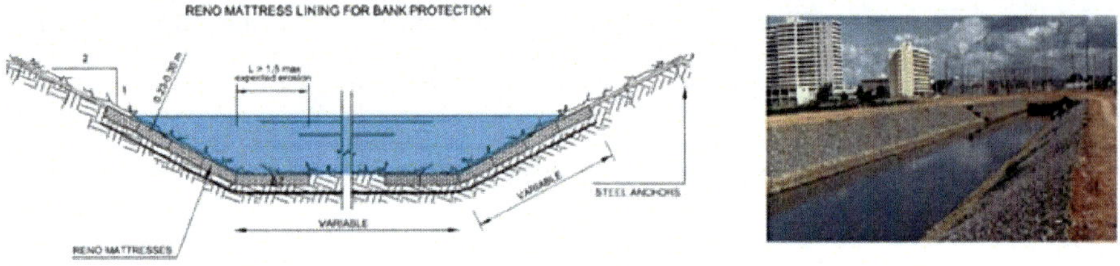

Fig. 8 Protection of drainage channel

to have excessive scour during low flow season. Usually, guide bunds are built with earth which requires erosion control on both sides.

2.4 Drainage Channels

In urban areas drainage channels, like channelized streams, collect and convey water to a safe zone. Any instability in these channels affects population and local infrastructure. Protection measures in these courses guarantee that the flow is safely brought to the delivery network. Figure 8 presents the typical protection adopted for drainage channels providing apron only requires length and not throughout the width.

2.5 River Training

In large floodplains, normally the groundwater flow, as well as the secondary channels may affect the flow in the main channel section sometimes, forcing changes in its path. Considering that floodplains (or areas adjacent to the main channel) are often used for agricultural purposes in the dry season, gabions and mattresses may be used to contain the flow inside a desired section.

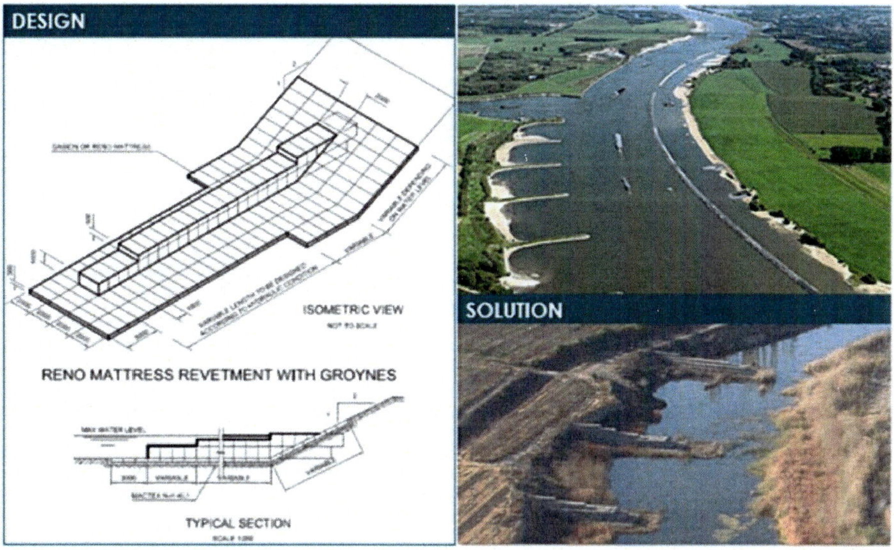

Fig. 9 River training using groyne

Often in wide river sections typically in alluvial plains, where meandering can be dangerous as it could progressively scour close to infrastructures groynes may be used to favor sedimentation in the outer reach of the bend. River training is the stabilization of the channel in order to maintain the desired cross section and alignment. River training in wide sections is usually carried out using groynes as indicated in Fig. 9.

2.6 *Lateral Spillways*

Lateral spillways are used to provide a controlled water release, diverting the water from the river during the flood, or redirecting the water back into the river, after the peak flood. Spillways release water to prevent damage to the levee embankment as presented in Fig. 10. Except during the high-water event, the water does not flow over the spillway. Spillways can be constructed with the crest lower than the top of the embankment; the protection is usually made by concrete, stones, rocks, or gabions. Overflow spillways with significant hydraulic heads require a properly shaped crest, an energy dissipator downstream, and a protected release channel.

3 Test Campaign and Results

The design practice for gabions and mattresses made of double twisted wire mesh as hydraulic structures for use erosion protection systems is based on a research

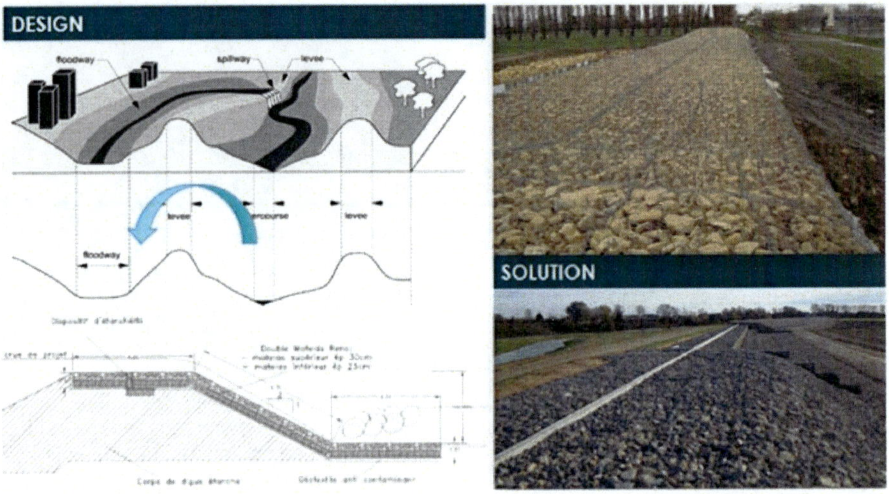

Fig. 10 Lateral spillway structure

carried out at the engineering research center of CSU (Colorado State University, Fort Collins—CO) reported by Simons et al. [11]. In the hydraulics of open channels, an important change over the years has been the creation of standards. The stability of erosion control systems depends on the ability to prevent soil loss underneath, as well as to maintain their integrity under the effects of the flow.

In 1984 the only design criterion for rock-based erosion systems in open channels was the theory of tractive force through the use of the Shields parameter. The mattress stability was based on the principle that the confined rock shall not move. However, the proliferation of several types of erosion protection systems required engineers to provide guidance in their selection, to guarantee effective protection against erosion. ASTM D6460 is the standard defining performance limits for erosion protection systems in open channels. This standard can be applied to all types of erosion protection systems as it defines the stability for a lining when either no collapse or erosion underneath occurs.

Thornton et al. [13, 14] an investigative study through a new research program was carried out at the engineering research center of CSU (Fig. 11). The objective of this research was to introduce a new design approach based on the observation of stability of mattress with respect to flow through mattress layer, stone movement inside mattress, and soil loss under the mattress. The new approach, in alignment with the established ASTM D 6460-19, was deemed more suitable for erosion protection systems, as it correlates stability to the soil loss underneath the lining. The research program was aimed at studying the improved limit performance of a new type of mattress Reno Mattress Plus (Fig. 12) versus conventional rock-filled mattresses in open channels.

The new mattress was tested on a 3′ × 30′ size flume where units were placed on a 9″ soil layer (Fig. 10). The performance against erosion was evaluated by assessing

Fig. 11 New mattress testing at CSU

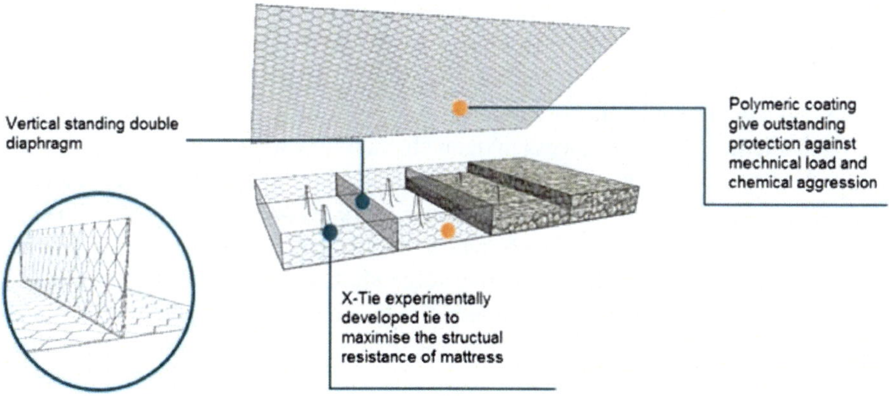

Fig. 12 New type of mattress

the effect of the stone motion inside the mattress, in relationship to its thickness and to the stone size, under multiple hydraulic flow regimes.

The study allowed to numerically evaluate the contribution of the different components of the new mattress, such as the type of partitions, the stone grading used for the filling and, the presence of vertical connections between the base and the lid (vertical ties). The combined effects of all these components led to the definition of improved overall performance with allowable shear stresses for the new mattress around 40–70% higher than a standard mattress.

The research resulted in a new design method for rock-filled mattresses which includes the effects of each component, namely the stone size and uniformity, the mattress thickness, the type of diaphragm partition, and vertical ties.

3.1 Design Criteria for Stability

Erosion follows the principle that instability starts when shear stresses become critical generating soil particles to get to incipient motion with the flow. Critical velocities will be dependent on the type of soil, the water depth, and the channel section. Critical shear stresses on the riverbed surface will be at the incipient motion equal to the allowable shear resistance which depends on the type of soil.

Two methods are commonly applied to determine channel is stable assuming boundaries in static equilibrium: (1) the permissible velocity approach and (2) the permissible tractive force (shear stress) approach. In permissible velocity approach, the channel is assumed to be stable if the mean velocity is lower than the maximum permissible velocity. The tractive force (boundary shear stress) approach focuses on stresses developed at the interface between flowing water and materials forming the channel boundary. By Chow's definition, permissible tractive force is the maximum unit tractive force that will not cause serious erosion of channel bed material from channel bed [2].

Permissible velocity approach was first developed around the 1920s. In the 1950s, permissible tractive force approach became recognized, based on research investigations conducted by the U.S. Bureau of Reclamation. Procedures for the design of vegetated channels using the permissible velocity approach were developed by the SCS and have remained in common use.

Considering physical process in open channel flow a more realistic model of soil detachment and erosion process is based on permissible tractive force which is the recommended method. The hydrodynamic force of water flowing in a channel is known as the tractive force. The basis for stable channel design with flexible lining materials is that flow-induced tractive force should not exceed the permissible or critical shear stress of the lining materials. In a uniform flow, the tractive force is equal to the effective component of the drag force acting on the body of water, parallel to the channel bottom [2].

The tractive force method has a couple of advantages compared to the permissible velocity method. First, the failure criteria for a given lining are represented by a single permissible shear stress value that is applicable over a wide range of channel slopes and channel shapes. Second, shear stresses are easily calculated using equations described in the next paragraph. The equations used are also useful in judging the field performance of a channel lining, because depth and gradient may be easier to measure in the field than channel velocity. The advantage of the permissible velocity approach is that most designers are familiar with velocity ranges and have a "feel" for acceptable conditions.

Tractive force method applied shear stresses. In an open channel, the flow exerts a force that acts on channel bed, following in the same direction of water flow, that's called tractive force or shear stress. The force scheme of the channel is shown in Fig. 13.

The acting forces are represented by the force of the water in the direction of the flow, upstream to the block, "$F1$", plus the block of water weight under analysis

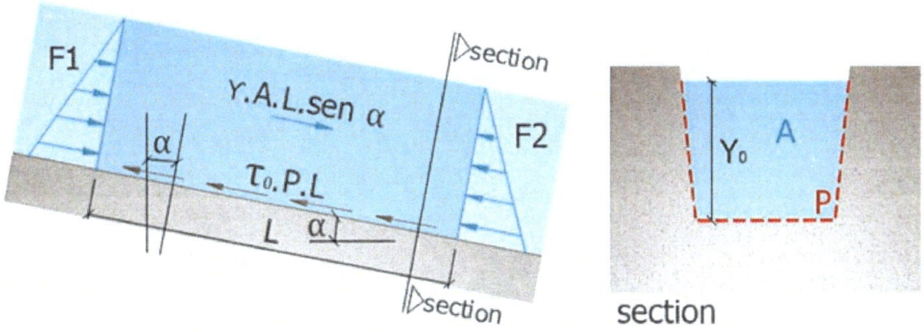

Fig. 13 Forces acting on the fluid mass (After Porto [10])

"$\gamma_w.A.L.\sin\alpha$", where "γ_w" is the specific weight of the water, "A" is the cross-sectional area, "L" is the length of the block under analysis and "α" is the longitudinal angle. The resistant forces are represented by the force of the water against the direction of flow, downstream to the block under analysis, "$F2$", plus the resistant stress that is occasioned by the bed "$\tau_0.P.L$", where "τ_0"is the shear stress and "P" is the wetted perimeter. The sum of all forces is represented on Eq. 1.

$$\sum F_x = F1 + \gamma_w.A.L.\sin\alpha - F2 - \tau_0.P.L \tag{1}$$

Considering that the system is in equilibrium it is possible to say that $F1 = F2$. For $a < 6°$, "$\sin\alpha$" is approximately equal to "$\tan\alpha$" that is approximately equal to the longitudinal slope of the channel "S". Adjusting the values, the shear stress can be calculated with Eq. 2.

$$\tau_0 = \gamma_w.R.S \tag{2}$$

Actual shear stress on the banks. Shear stress in channels is not uniformly distributed along the wetted perimeter [2, 9]. A typical distribution of shear stress in a prismatic channel is shown in Fig. 14. The shear stress is zero at the water surface and reaches a maximum on the centerline of the channel. The maximum shear stress on the side slopes occurs at about the lower third of the side. In the example of Fig. 15, the maximum shear stress on the bottom is 97% of "τ_0". As this percentage is generally close to 100%, "τ_0" can be always be adopted without being reduced.

On the other side, generally, the maximum shear stress on the banks is 76% or less of "τ_0". In order to determine this value, the reduction coefficient (K_m) can be obtained from Fig. 16 according to the section geometry (where "Z" is in relation to the slope, H:Z; V:1).

Maximum shear stress on the banks can be determined using the Eq. 3.

$$\tau_m = \tau_0.K_m \tag{3}$$

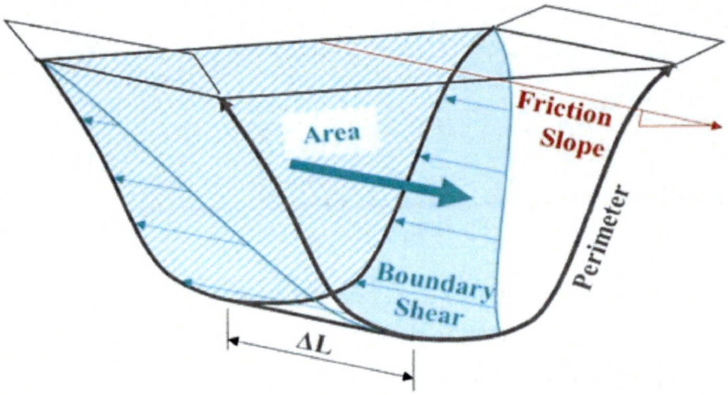

Fig. 14 Example of shear stress distribution in a trapezoidal channel section (After Chow [2])

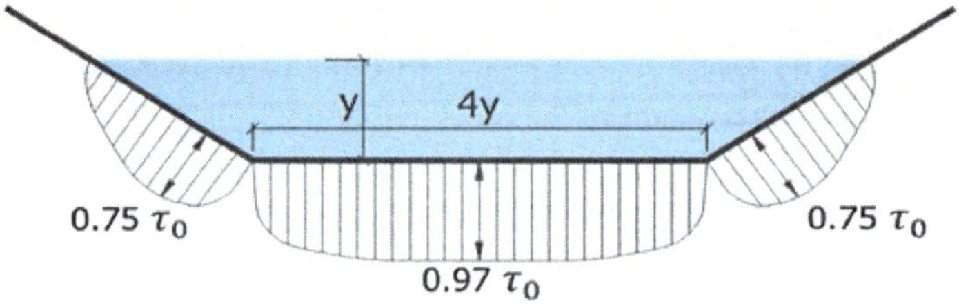

Fig. 15 Example of shear stress distribution in a trapezoidal channel section (After Chow [2])

Fig. 16 Shear stress on the sides of the channel (After Olsen and Florey [9] and Lane [7])

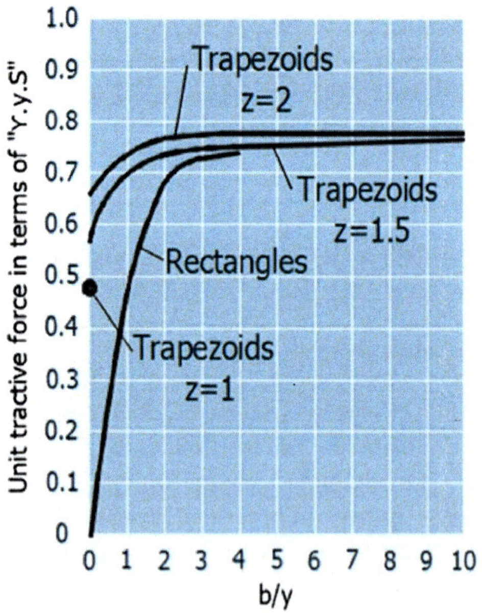

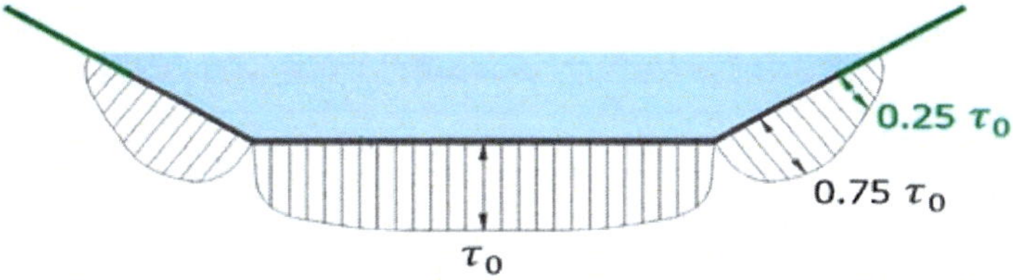

Fig. 17 Example of shear stress coefficient according to the height of the revetments

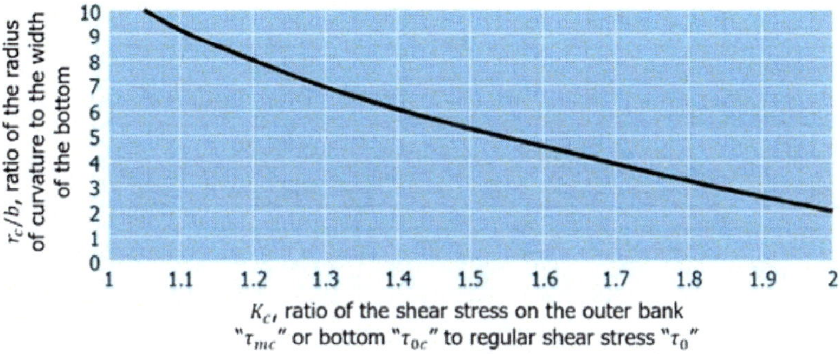

Fig. 18 Shear stress increase on curved channel (After S.C.S. [12])

New shear stress coefficient can be estimated based on the water depth at the point of contact of the section. For example, Fig. 17 presents the maximum shear stress over the green stretch is equal to 25% of "τ_0". This variation of shear stress mentioned by authors [7, 9].

Shear stress on channel bends. Curved channels have higher shear stresses on the outer banks than in straight channels. Figure 18 present an indicative relation between the increase factor and the ratio between curvature and channel width. The increased shear stresses on the banks and on the bottom are calculated respectively, using Eqs. 4 and 5.

$$\tau_{\mathrm{mc}} = \tau_{\mathrm{m}}.K_{\mathrm{c}} \tag{4}$$

$$\tau_{0\mathrm{c}} = \tau_0.K_{\mathrm{c}} \tag{5}$$

where "τ_{mc}" is the acting shear stress on the external side of the curved channel, and "$\tau_{0\mathrm{c}}$" is the acting shear stress on the bottom of the curved channel.

Allowable shear stress of soils or lining materials. The acting shear stresses shall be compared with the allowable shear stress of the materials which are presented in Table 1 [7].

The critical shear stress of Riprap (loose rock) can be calculated using the Shields Eq. (6).

Table 1 Allowable shear stress of soils and lining materials

Material	Allowable shear stress (N/m^2)
Natural soils	
Fine sand	3.50
Sand and gravel	15.30
Coarse gravel	32.00
Cobbles and shingles	52.60
Stiff clay (cohesive)	22.00
Shales (cohesive)	32.00
Silts w/cobbles (cohesive)	38.00
Soil bioengineering	
Grass mats	10.00
Cutting shrubs	10.00
Brush mats w/willow	50.00
Riparian wattles	10.00
Willow protections	20.00
Natural vegetation	
Dense grass and light bushes	30.00
High bushes	60.00
Dense vegetation with arboreous trees	300.00
Dense bushes	50.00
Low arboreous plants	100.00

$$\tau_c = C^*.(\gamma_S - \gamma_w).d_m \tag{6}$$

where "C^*" is the Shields parameter, for rip rap equal to 0.047 as per Kilgore and Cotton [5], and "γ_S" is the stone unit weight (the minimum thickness of the riprap lining should be the greatest between the D_{100} and $1.5 * D_{50}$). d_m is mean particle size.

Allowable shear stress of gabions and reno mattresses plus. Thornton et al. [13] ran new tests with new mattresses to investigate the influence of mesh, uniformity coefficient, and average diameter of filling rock on flow dissipation. New mattresses are manufactured with 1 m spaced double-sided (pleated) diaphragms and have 1 X-Tie per m^2 (Fig. 12).

Mattresses of different thicknesses with X-ties were tested. The research led to introduce partial correction factors to Shield's equation (Eq. 7) to obtain the allowable shear stress of Reno mattresses plus and extrapolated for gabions as well (Table 2).

$$\tau_a = C^*.(\gamma_S - \gamma_w).d_m.k_{D50}.k_{Cu}.k_{thickness}.k_{castoro}.k_{ties} \tag{7}$$

Table 2 Allowable shear stress of Reno mattress plus and gabion

Thickness (m)	D_{50} (m)	Uniformity coefficient (Cu = D_{60}/D_{10})	Allowable shear stress (N/m^2)
0.17	0.083	1.00	402
0.17	0.083	1.50	321
0.17	0.095	1.00	445
0.17	0.095	1.50	355
0.23	0.083	1.00	482
0.23	0.083	1.50	385
0.23	0.095	1.00	534
0.23	0.095	1.50	426
0.23	0.102	1.50	445
0.30	0.083	1.00	576
0.30	0.083	1.50	459
0.30	0.095	1.00	638
0.30	0.095	1.50	509
0.30	0.102	1.50	532
0.30	0.121	1.50	591
0.50	0.127	1.50	700

where the partial factors are: "k_{D50}" related to flow rate due to average diameter, "k_{Cu}" related to flow rate due to uniformity coefficient, "$k_{thickness}$" due to mattress thickness variation, "$k_{castoro}$" due to the type of mattress partition and "k_{ties}" due to the use or not of X-Ties.

4 Case Studies

4.1 Mahanadi and Devi Nadi Orrisa (Near Paradeep)

The embankment and riverbed of Devi river was severely eroded in October 1999 cyclone. Mahanadi river near Paradeep also was severely eroded with large pocket formations. Some temporary protection works (riprap and bamboo piling) were observed to be dislodged. Devi river sites being close to Bay of Bengal, small tidal waves (in the range of 0.2–0.5 m) are present throughout the year. These waves combined with the velocity of water (2–3 m/s) have dislodged the existing protection works. The project was a World Bank aided project. The riverbanks which were at a very precarious and unstable slope were reconstructed. The embankment was reconstructed using locally available earth fill compacted to 90% of maximum dry density. The slope was stabilized by loose dumping of rocks (rip rap) 30–40 kg weight to form a bank slope of 1:2. Permanent protection of the formed bank slope was carried out with 0.5 m thick Gabion revetment (Figs. 19 and 20). The project involved underwater placement of Gabion mattress. The prefilled mattresses were placed on the bed of the river with a crane. The mattresses were placed at a depth

Fig. 19 Protection work on river Mahanadi

Fig. 20 Protection work on river Mahanadi after 10 years

of 15–18 m under water. Flexible mattresses adopted to the bank slope were cost effective.

4.2 *Delhi Noida Bridge Across the River Yamuna (1999)*

The Delhi Administration had proposed a road bridge across the river Yamuna downstream of the existing ISBT (Inter-State Bus Terminus) road bridge, for the Mass Rapid Transport System. The proposed bridge was about 2 km upstream of the existing Okhla barrage and had a waterway of 554.4 m. Because of the bridge, it was anticipated that the velocity would be very high and hence the piers of the bridge and the bed had to be protected. The protection works for the river training works were designed for a design discharge of 14,866 cum/s and a maximum velocity of 3.3 m/s. The original design of the river training works all provided for massive apron and slope protection, using 1 m thick stones by way of pitching. The original tendered estimate was about INR 275 million. It was proposed to use a single layer of 0.5 m thick gabion mattress. Zinc-coated gabions of double twisted mesh and wire dia 3.0 mm were used. This resulted in a reduction of apron thickness by 50% as compared to conventional protection and resulted in savings in construction costs (nearly INR 130 million). Figures 21 and 22 presents a view of the bed lining provided at Delhi Noida bridge immediately after construction and after 20 years respectively.

Fig. 21 Guide bund protection at Delhi Noida bridge 1999

Fig. 22 Guide bund protection at Delhi Noida bridge after 20 years

4.3 River Training Work Near Guide Bund on River Gandak

Bihar Rajya Pul Nirman Nigam limited was constructing a bridge on river Gandak. Due to limited river opening and sudden increase in discharge, the river flow started eroding Gopalganj side bank. On the banks and toes of the downstream guide bund, tension cracks were visible. The stability of constructed guide bund and the areas lying downstream towards Gopalganj were highly susceptible to be washed off and could have triggered possible channel avulsion. Riverbank protection measures were urgently required. Protection work is required to be installed as soon as possible and under mercy water conditions. It was required to use available sand and transportation of stones to be avoided. Protection work in the form of strong, mechanically selvedge, double twisted gabions lined with geotextile and filled with locally available soil was suggested (Fig. 23). These gabions are factory assembled and supplied in folded form. Construction demanded quick formation of gabion and to be filled faster for which special type of arrangement was designed. The protection was terminated with bullhead structure near the end to divert the flow.

4.4 Strengthening Work of Existing River Embankment at River Hooghly, Budge Budge, South 24 C Parganas

The Cheviot group of companies is located along the river Hooghly, at Budge Budge, 24 Parganas, Kolkata, and West Bengal. Tidal Bore phenomenon occurs in the estuary zone and the intense turbulence and the hydraulic jump, caused as result had led to damage of the compound wall of Cheviot jute mill. Maximum height of the tidal bore observed was approximately 2.10 m and the velocity was 7–10 m/s. Tidal Bore occurred once or twice a month and the condition extended for a period of 2–6 days occurring once each day.

The height of the embankment to be protected was approx. 6 m and slope 1V:2H. The thickness of the system was analyzed and considered as approximately 3 m.

Fig. 23 Protection work with sack gabion filled with sand at Gopalganj

Geotextile bags and sack gabions filled with sand was adopted as a unit for solution system. These gags were produced from woven polyester geotextile material. Sack gabions lined with geotextile material containment system were filled with dry sand and placed in position. These were used for the preparation of the foundation base for the structures. The erosion measures constructed with geotextile bags, sack gabions, double twist mesh netting, mattress effectively functioned protected the banks, against the tidal bore (Fig. 24).

4.5 Protection Work on Left Bank of Gandak Barrage at Valmikinagar

Water resources department of Bihar was looking for protection of the left bank of Gandak barrage. Gandak river received a high flow of water of approximate discharge 24,100 cumecs during monsoons which result in severe erosion on the left river bank endangering stability. The deposition of eroded bank material further towards downstream can damage the gates of Gandak barrage. Approximately, 1080 m length of the bank was required to be protected.

The retaining structure was designed for seismic zone V as per IS 1893. The retaining structure is required to be flexible, permeable in nature to withstand the pore water pressure and resistant to scour. In order to serve all these functions, the entire solution was divided into three parts: (i) Reinforced soil retaining wall with

Fig. 24 Protection work with mattress at budge budge

permeable gabion facia of Terramesh made up of double twisted steel wire mesh, (ii) surficial protection works with synthetic erosion control mat reinforced with double twisted steel wire mesh and (iii) scour protection in the form of launching apron made up of mattress (Fig. 25). A gabion toe wall was provided at the end of the apron. The solutions were chosen because of their high permeability, flexibility, and environment friendly.

5 Summary

Erosion and sedimentation in watercourses are inevitable processes and it is required to understand their nature. Erosion control using sustainable systems like gabions, mattresses, etc. which will yield and recover the damaged environment are presented. Considering physical process in open channel flow a more realistic model of soil detachment and erosion process is based on permissible tractive force (shear stress) which is the recommended method. The tractive force method has advantages compared to the permissible velocity method. The failure criteria for a given lining are represented by a single permissible shear stress value that is applicable over a wide range of channel slopes and channel shapes.

The first experimental studies on the stability of mattresses were in line with the current design method used for riprap, based on the tractive force theory. This method is focused on the mechanical stability of the lining (condition of no movement of

Fig. 25 Reinforced soil wall, launching apron, and erosion control mat at Valmiki Nagar

the stone), without consideration of the potential erosive effects generated under the lining. The latest test campaign based on ASTM D6460 introduced a new way of looking at the stability of erosion protection systems: mechanical stability combined with a controlled (or limited) erosion underneath. Double twisted wire mesh products have been used in varied environmental conditions in India. The solutions are performing the intended purpose to be cost effective and environmentally friendly.

References

1. ASTM D 6460 (2019) Standard test method for determination of rolled erosion control product (RECP) performance in protecting earthen channels from stormwater-induced erosion
2. Chow VT (1959) Open-channel hydraulics. McGraw-Hill, Tokyo
3. Hjulstrom F (1935) Studies of the morphological activity of rivers as illustrated by the River Fyris, vol 25. Bulletin of Geological Institution, University of Uppsala, Uppsala
4. IS 1893, Part 1 (2016) Criteria for earthquake resistant design of structures—Part 1: general provisions and buildings
5. Kilgore RT, Cotton GK (2005) Design of roadside channels with flexible linings HEC No. 15, 3rd edn. NHI, U.S. Department of Transportation, Federal Highway Administration, Fort Arlington
6. Lagasse PF, Clopper PE, Pagán-Ortiz JE, Zevenbergen LW, Arneson LA, Schall JD, Girard LG (2009) Bridge scour and stream instability countermeasures—experience, selection and design

guidance. HEC No. 23, 3rd edn. NHI, U.S. Department of Transportation, Federal Highway Administration, Fort Arlington

7. Lane EW (1952) Progress report on results of studies on design of stable channels—Hyd-352. U.S. Department of the Interior, Denver
8. Natural Resources Conservation Service (2007) National engineering handbook—part 654. Engineering Division, U.S. Department of Agriculture, Washington D.C.
9. Olsen OJ, Florey QL (1952) Sedimentation studies in open channels: boundary shear and velocity distribution by membrane analogy, analytical and finite difference method. USBR, Lab. Report No. SP-34
10. Porto RM (2006) Hidráulica básica, 4ª Ed. EESC-USP, São Carlos
11. Simons DB, Chen YH, Swenson LJ (1984) Hydraulic test to develop design criteria for the use of Reno mattress. Civil Engineering Department, Engineering Research Center, Colorado State University
12. Soil Conservation Service (1977) Design of open channels, Technical Release N° 25. Engineering Division, U.S. Department of Agriculture, Washington D.C.
13. Thornton CI, Hogan T, Keene A, Eggleston S (2019) Hydraulic testing and data report for phase 1—rock gabion mattress testing. Daryl B. Simons Building—Engineering Research Center, Colorado State University, Fort Collins
14. Thornton CI, Hogan T, Pietro PD, Mandavkar S, Mazzon N (2020) Experimental study on revet mattresses in open channel flow: a new design approach against erosion. Colorado State University and Maccaferri

Geotextile Tube for Coastal Protection and Land Reclamation

Albert L. K. Lim and K. H. Siew

Abstract Geotextile tube is used to encapsulate sandy soils to enable their use as flexible, erosion-resistant, mass-gravity structures in coastal protection and land reclamation applications. The paper introduces the various grade of geotextile tube units in use and discusses their range of applications. Important stability aspects of design are presented along with geotextile tube mechanical and hydraulic property requirements. Ways of enhancing geotextile tube durability are also discussed. Two case studies are presented that demonstrate the effective use of geotextile tubes for coastal protection and land reclamation project.

Keywords Geotextile tube · Woven · Composite · Coastal protection · Land reclamation

1 Geotextile Tubes for Coastal Engineering Applications

1.1 Engineering Features of Geotextile Tubes

Geotextile tubes are laid out and filled hydraulically onsite to their required geometrical form. The typical features of a geotextile tube are shown in Fig. 1. Hydraulic fill is pumped into the geotextile tube through specially manufactured filling ports located at specific intervals along the top of the geotextile tube. During filling, the geotextile tube, being permeable, allows the excess water to pass through the geotextile skin while the retained fill attains a compacted, stable mass within the tube. For hydraulic and marine applications, the type of fill used is sand or a significant percentage of sand. The reasons for this are that this type of fill can be placed to a good density by hydraulic means; this type of fill has good internal shear strength; and this type

A. L. K. Lim · K. H. Siew (✉)
TenCate Geosynthetics Asia Sdn Bhd, Shah Alam, Malaysia
e-mail: kh.siew@tencategeo.com

A. L. K. Lim
e-mail: lk.lim@tencategeo.com

C. N. V. S. Reddy and S. Sassa (eds.), *Scour and Erosion Related Issues*,
Lecture Notes in Civil Engineering 177,
https://doi.org/10.1007/978-981-16-4783-3_7

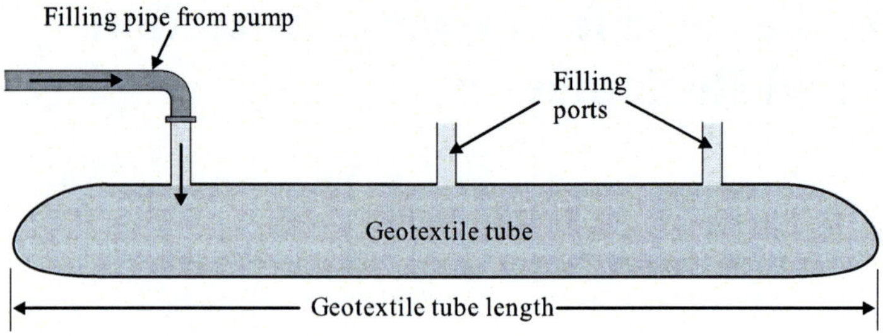

Fig. 1 Typical features of geotextile tubes

of fill, once placed, will not undergo further consolidation, which would change the filled shape of the geotextile tube. Once filled, the geotextile tube behaves as a mass-gravity unit and can be designed accordingly.

The geotextile tube skin performs three functions that are critical to the performance of the filled geotextile tube. First, the geotextile skin must have the required tensile strength and stiffness to resist the mechanical stresses applied during filling and throughout the life of the units and must not continue to deform so that the geotextile tube changes shape over time. Second, the geotextile skin must have the required hydraulic properties to retain the sand fill and prevent erosion under a variety of hydraulic conditions. Third, the geotextile skin must have the required durability to remain intact over the design life of the units.

Geotextile tubes are normally described in terms of either a theoretical diameter, *D* (in Europe, Middle East and Asia) or a circumference, *C* (in North and South America), Fig. 2a. While these two properties represent the fundamental parameters of geotextile tubes they are not of direct interest when it comes to the engineering parameters for hydraulic and coastal applications where the geotextile tube in its filled condition is of prime importance. The various engineering parameters of importance are shown in Fig. 2b.

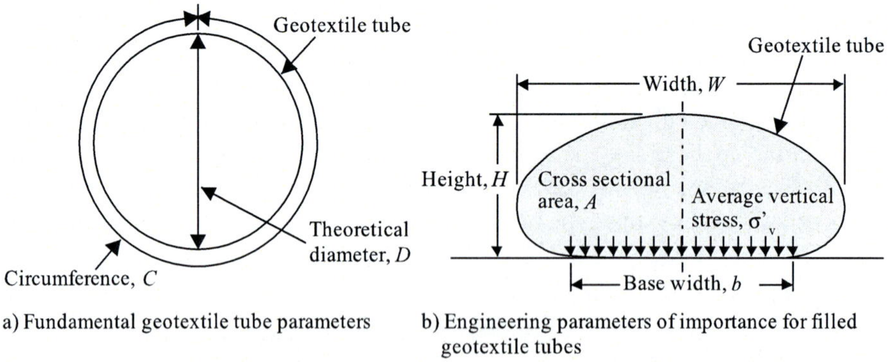

Fig. 2 Various parameters associated with geotextile tubes

Table 1 Approximate relationships between fundamental and engineering parameters of geotextile tubes (After Lawson [1])

Engineering parameter	In terms of theoretical diameter, D	In terms of circumference, C
Maximum filled height, H	$H \approx 0.55D$	$H \approx 0.18C$
Filled width, W	$W \approx 1.5D$	$W \approx 0.5C$
Base contact width, b	$b \approx D$	$b \approx 0.3C$
Cross sectional area, A	$A \approx 0.6D^2$	$A \approx 0.06C^2$
Average vertical stress at base, σ_v'	$\sigma_v' \approx 0.7\gamma D$	$\sigma_v' \approx 0.22\gamma C$

Note γ = bulk density of the geotextile tube fill

Table 1 lists approximate relationships between the fundamental geotextile tube parameters of theoretical diameter and circumference (Fig. 2a) and the engineering parameters of importance depicted in Fig. 2b. The relationships are applicable to geotextile tubes that have a maximum strain $\leq 15\%$, low unconfined creep and are filled to maximum capacity with sand-type fill. Furthermore, it is also assumed that the foundation beneath the tube is a flat, solid surface.

1.2 Applications for Geotextile Tubes

Geotextile tubes are used for a range of hydraulic and coastal applications where mass-gravity barrier-type structures are required. These applications are shown in Fig. 3 and described briefly below.

Geotextile tubes are used for revetment structures where their contained fill is used to provide mass-gravity stability, Fig. 3a. They are used for both submerged as well as exposed revetments. For submerged revetments, the geotextile tube is covered by local soil and is only required to provide protection when the soil cover has been eroded during periods of intermittent storm activity. Once the storm is over the revetment is covered by soil again either naturally or by maintenance filling. For exposed revetments, the geotextile tube is exposed throughout its required design life.

To prevent erosion of the foundation soil in the vicinity of the geotextile tube it is common practice to install a scour apron (see Fig. 3a). This scour apron usually consists of a geotextile filter layer that passes beneath the geotextile tube and is anchored at the extremity by a smaller, filled, geotextile tube.

Revetments are also constructed using multiple-height geotextile tubes. Here the geotextile tubes are staggered horizontally to achieve the required stability. Considerable care should be exercised during the construction of these types of revetments to ensure the water emanating from the hydraulic filling of the upper geotextile tubes

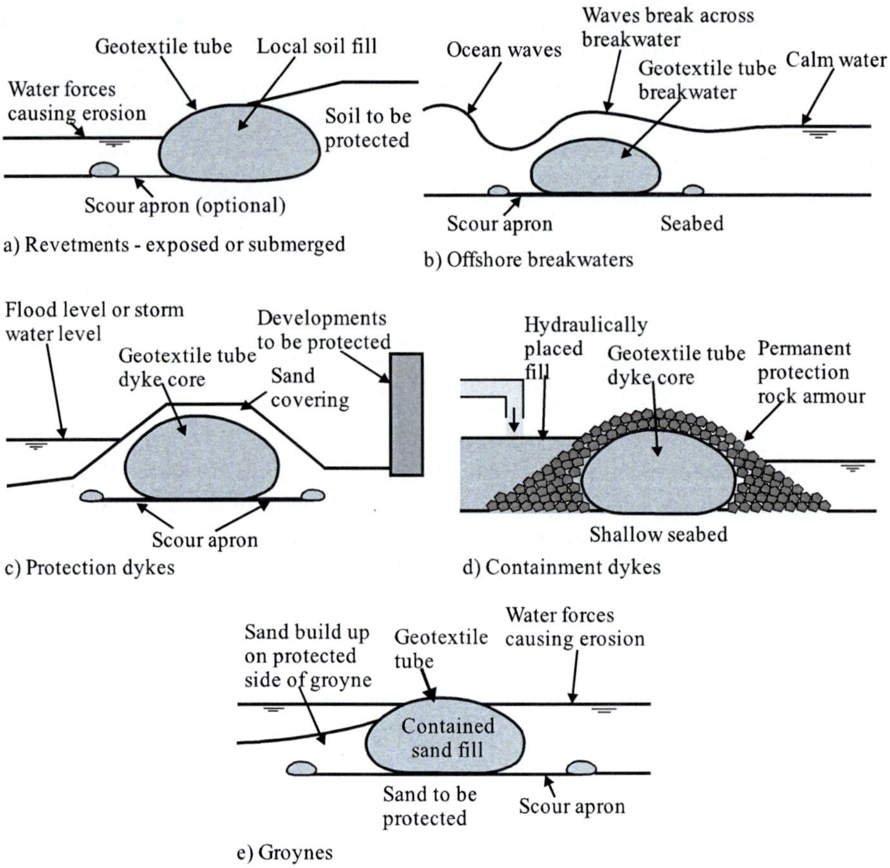

Fig. 3 Hydraulic and coastal applications for geotextile tubes

does not erode the soil and undermine the lower geotextile tubes in the multiple-height revetment structure. Examples of use are given by Nickels and Heerten [2] and Artières [3].

Geotextile tubes are used for offshore breakwaters to prevent the erosion of shoreline developments, Fig. 3b. Here the filled geotextile tube is located a certain distance offshore in order to dissipate wave forces before they can reach the shoreline. Again, scour aprons are used beneath the geotextile tube breakwater to ensure local erosion does not undermine the breakwater structure.

Geotextile tubes are used for protection dykes where they prevent flood and storm damage to valuable structures and real estate, Fig. 3c. Protection dykes also may be used for river, lake or stream training works.

Where geotextile tube protection dykes are constructed it is common to cover the geotextile tube with local soil. The geotextile tube is only required to function intermittently during storm or flood periods when the soil cover is eroded. The use of the soil cover provides a number of advantages to the geotextile tube core. First, the soil cover hides the geotextile tube core thereby providing an aesthetic environment and ensuring no damage due to vandalism. Second, the soil cover protects the geotextile tube from long term exposure to the atmosphere (UV degradation).

Where geotextile tubes are used for river, lake or stream training works it is common to leave the tube exposed except for major structures where rock armour layers may be placed over the geotextile tube to dissipate hydraulic forces. In this type of application, composite geotextile tube which has the characteristics of sand entrapment and high UV resistance are used.

Geotextile tubes are used for the cores of containment dykes where water depths are relatively shallow, Fig. 3d. Here, the tube structure contains a filled reclamation area—the reclamation fill being dry dumped or placed hydraulically. The advantage of this approach is that the same hydraulic fill used in the reclamation can also be used inside the geotextile tubes for the containment dykes thus avoiding the need to import rock fill for the dykes. Where water forces dictate and where longevity is required, rock armouring can be placed around the geotextile tube core, e.g. Figure 3d. Examples of use are given by de Bruin and Loos [4], Spelt [5], Fowler et al. [6] and Yee [7].

Geotextile tubes can be used as groynes to prevent the littoral movement of sediment, Fig. 3e. In most cases the geotextile tubes are left exposed, but coatings or a rock covering may be applied depending on the circumstances and the required life expectancy. Examples of use are given by Jackson [8] and Fowler et al. [6].

1.3 Design Procedure for Geotextile Tubes

Since geotextile tubes behave as mass-gravity units the conventional approach to design follows a standard procedure of assessing the possible modes of failure or deformation in order to arrive at a safe design solution. Figure 4 lists the various limit state modes that should be assessed and these are divided into external modes (those modes affecting the performance of the geotextile tube structure overall) and internal modes (those modes affecting the performance of the internal structure of individual geotextile tubes). Either a global factor of safety or a partial factor of safety approach can be applied when assessing the various limit state modes.

There are six external limit state modes to be assessed, Fig. 4a. These are sliding resistance, overturning resistance, bearing resistance, global stability, scour resistance and foundation settlement. Geotextile tubes are very stable units with high base contact width to height ratios, e.g. in Table 2 $b/H = 1.5$. Geotextile tubes should be checked for sliding and overturning stability, especially if the tubes are of small theoretical diameter, $D \leq 2$ m. Bearing stability [e.g. Fig. 4a(iii)] may be of importance if the foundation is very soft and the geotextile tube is very large. However, experience has shown that the distribution of weight of geotextile tubes on soft foundation soils is very efficient.

Global stability only needs to be taken into account when multiple geotextile tubes are used, e.g. Fig. 4a(iv). Here, the stability analysis should take into account changes

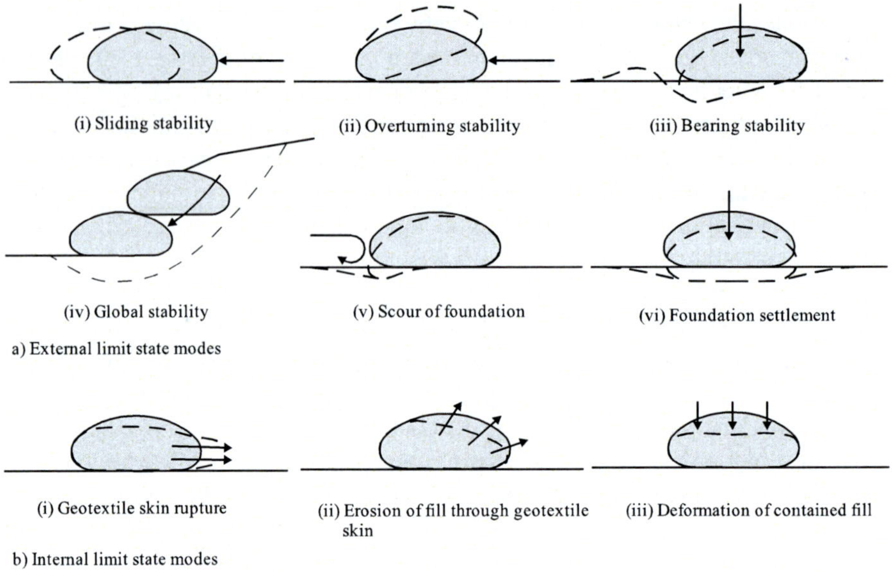

Fig. 4 Limit state modes for the design of geotextile tubes

Table 2 Technical properties comparison between TD value and 5 years exposure under Vietnam weathering condition value

Properties	Test standard	Unit	TD value	5 years exposed value	Min. retained (%)
Tensile strength, MD	ISO 10319	kN/m	200	112.9	56.5
Tensile strength, CD	ISO 10319	kN/m	200	130.1	65.1
CBR puncture strength	ISO 12236	kN	22	13.7	62.3
Pore size O90	ISO 12956	mm	0.35	0.3	100.0

in both the external water level and the groundwater level within the geotextile tube structure. Also, potential weak planes between adjacent geotextile tubes should be assessed.

Scour of the foundation around the edges of geotextile tubes [e.g. Fig. 4a(v)] can lead to undermining and the geotextile tube overturning. Scour may occur either during the filling process or during the life of the tube. During filling, a large amount of water is expelled through the geotextile skin and this can cause erosion and undermining of the geotextile tube if measures are not taken to prevent this. To prevent scouring of the foundation during filling it is common practice to first install a geotextile or geomembrane layer beneath the geotextile tube prior to tube placement and filling. This procedure is very important where multiple-height geotextile tubes are installed in order to prevent the filling water of the upper tubes causing erosion and instability of the lower tubes in the structure.

Where there is potential for foundation scour during the life of the geotextile tube structure it is common practice to install a scour apron during construction, see Fig. 3. The scour apron consists of a geotextile filter anchored at the extremities by means of a small diameter geotextile tube manufactured as an integral part of the geotextile filter base.

Where geotextile tubes are constructed on compressible foundations and where they are required to meet specific height requirements for hydraulic structures (e.g. breakwaters), an assessment of the effect of foundation settlement should be performed, Fig. 4a(vi).

There are three internal stability modes to be assessed, Fig. 4b. These are geotextile skin rupture resistance, geotextile skin hydraulic resistance and deformation of the contained fill. Geotextile skin rupture resistance is discussed in more detail below.

1.4 Required Tensile Properties of Geotextile Tubes

During the filling process and throughout the life of filled geotextile tubes tensions are generated in the tube unit. These locations are around the circumference of the geotextile tube, along the length, or axis, of the geotextile tube and at the connection of the filling ports with the geotextile tube.

The analysis of tensions generated in geotextile tubes is complicated due to the effect of geotextile tube geometry. Further, the fill contained within geotextile tubes starts as a liquid, i.e. with zero shear strength and then fairly quickly reverts to a solid, i.e. with internal shear strength. This change in phase of the contained fill, the amount of filling and pumping pressure applied and the time over which the contained fill changes in phase all affect the magnitudes of the tensions generated in geotextile tubes. For hydraulic and coastal structures where the contained fill consists of sand, the time it takes to change to a solid material is very short (unlike finer fills) and thus analysis methods based on the assumption of a shear resistant fill are more appropriate for this type of application.

The procedure normally used to determine the tensions in geotextile tubes is to first determine the circumferential tension, then the axial tension and finally the port connection tension. Figure 5 shows the maximum circumferential tension $[T_{max}]c$ for filled geotextile tubes having theoretical diameters $D = 3.0$ m, 4.0 m and 5.0 m using the procedure of Palmerton [9]. As noted in Table 1 the maximum filling height is $H \approx 0.55D$, which results in maximum circumferential tensions $[T_{max}]c$ of 18 kN/m, 30 kN/m and 50 kN/m respectively for the three geotextile tube sizes.

The geotextile tube skin and its component parts must have adequate tensile strengths to resist the tensions generated during the filling process and throughout the life of the filled geotextile tube. This tensile strength requirement must account for the tensions generated in the geotextile fabric itself as well as any seams used to join the geotextile fabric together. To arrive at safe tensile strengths a suitable factor

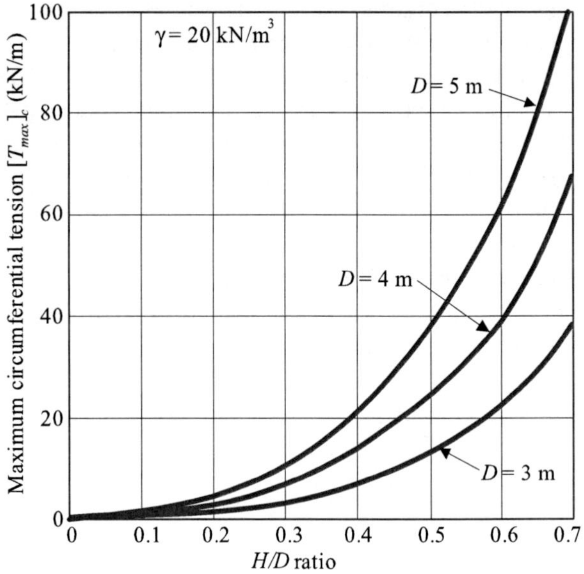

Fig. 5 Maximum circumferential tensions in geotextile tubes according to Palmerton [9]

of safety must be applied to the magnitude of the tensions generated at the various locations in the filled geotextile tube. Unless a specific analysis is undertaken of the various factors that comprise the overall factor of safety a default value of 4.0–5.0 is normally applied.

1.5 Protection Measures Applied to Geotextile Tubes

External protection measures are applied to geotextile tubes for a variety of reasons, namely;

- to reduce the impact of the hydraulic forces acting directly on the geotextile tube;
- to enhance the design life of the geotextile tube in an exposed environment;
- to protect from extreme natural occurrences, e.g. ice and debris flows, etc.;
- to protect from vandalism.

In many instances, geotextile tubes are required to perform over a relatively long design life in an exposed environment. In this environment ultra violet light (UV) degradation can occur, with the geotextile tube design life dependent on the level of UV radiation and the resistance of the geotextile tube skin to this radiation. If the geotextile tube is located in a marine environment, marine growth generally occurs quickly on the outer surface and this tends to mask the geotextile skin somewhat from the effects of UV radiation. However, for good long term performance in an exposed environment, additional protection measures are normally required for the geotextile tube skin. These measures are listed below in order of providing long term performance.

- Additional stabilizer packages in the geotextile tube skin—where the enhanced performance of the stabilizer package improves the performance of the geotextile tube skin over time.
- More robust composite geotextile skin—where extra design life is achieved by the use of more robust composite geotextile skins that degrade over a longer period of time.
- Soil covering—where the geotextile tube is covered by soil or sand to prevent long term UV exposure. Here the geotextile tube structure performs intermittently during periods of storm activity and is then covered over again by soil or sand.
- Armour covering—where a flexible armour covering is used around the geotextile tube structure to prevent long term exposure to UV light. This is normally used in hydraulic and marine applications where severe hydraulic forces occur.

2 High Strength Woven Engineering Geotextile Tube for Construction of Deep C Industrial Zones, Vietnam

2.1 Project Overview

Deep C was established in the year 1997 by a consortium of international investors led by the Belgian company Rent-A-Port and the Vietnamese local governments. Since then, it had evolved to be one of the biggest industrial zone developers in Vietnam with a total of 5 industrial zones and about 3000 ha of land for lease. The design for the land reclamation required a containment dyke to be constructed as a perimeter bund wall to encompass the area to be reclaimed.

The original consultant design of the containment dyke consists of a conventional armour rocks bund wall (Fig. 6) which is difficult to construct due to the soft underlying marine clay with an undrained shear strength of 5 kPa for the first 6–9 m and 10 kPa for the following 6–8 m.

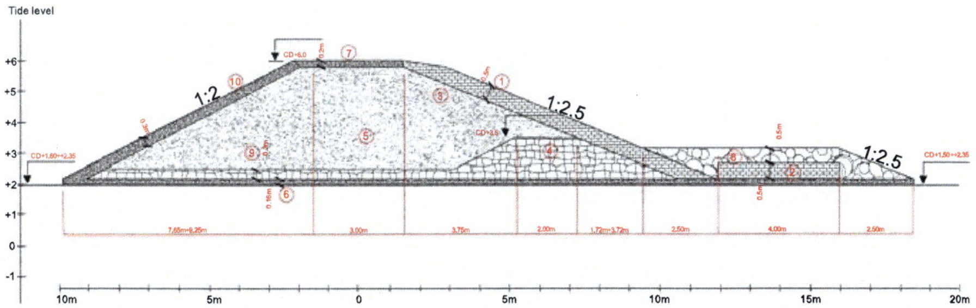

Fig. 6 Original design using conventional armour rocks system

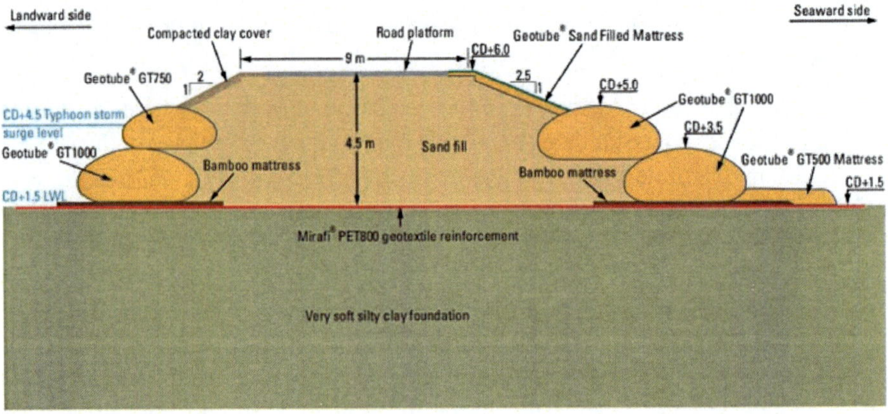

Fig. 7 Revised design using geosystem containment dyke

The successful use of Geosystem containment dyke system (Fig. 7) for the Lach Huyen bridge project which is adjacent to Deep C and Incheon Bridge project in Korea [12, 13] prompted the project owner to evaluate their original design to a Geosystem containment system. The Geosystem containment system is a combination of geotextile tube, PET high strength basal reinforcement and sand filled mattress and is cost effective, environmentally friendly and reduce construction time. In total 19 km of geotextile tube, 88,000 m^2 of basal reinforcement and 46,000 m^2 of sand filled mattress was successfully installed in this portion of the works.

2.2 *Site Conditions*

The subsoil in the area generally consists of very thick alluvial and marine clay deposits above a dense to very dense sand/gravel foundation. The soft clay layers have low bearing capacity and excessive settlement characteristics. The first layer consists of very soft organic sandy silty clay with thickness of about 6–9 m and undrained shear strength of about 5 kN/m^2. The second layer consists of soft to very soft clay with fine sand with thickness of about 6–8 m and undrained shear strength of about 10 kN/m^2. Below that is an approximately 4.5 m layer of stiff to very stiff clay with gravel. The fourth layer generally consists of medium stiff clay (Fig. 8).

2.3 *Design and Stability Analysis*

Conditions that influence the properties of the geotextile over time should be considered. The polymer used for the manufacture of geotextile tubes should be durable in a

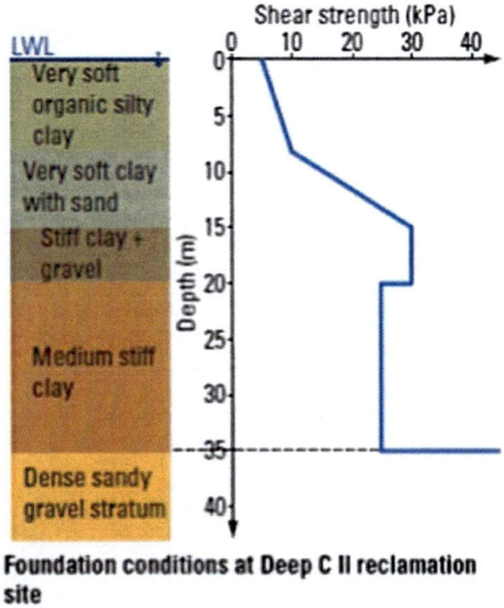

Fig. 8 Underlying soil properties at project site

biological, chemical environment and ultra violet light resistance. In the design analysis, a global factor of safety of 3.5–5 was applied for creep, construction damage, environmental damage, seam efficiency, etc. The geotextile tensile stresses of the tube during hydraulic filling were analyze using Geotube® Simulator software program; which is a computer program developed by TenCate Geosynthetics North America. The required ultimate circumferential and axial tensile strength of the tube were determined to be 115 kN/m and 89 kN/m respectively. As the design life of the geotextile tube needs to be 5 years in tidal conditions, a woven polypropylene geotextile tube of circumference 12.6 m with a tensile strength of 200 kN/m in both warp and weft directions and seam strength of 160 kN/m was supplied to the project. Exhumation of geotextile tube in Lach Huyen bridge after 5 years exposure in similar site condition has showed tensile strength retention of 50%. Table 2 lists technical properties comparison of the supplied woven polypropylene geotextile tube between technical data value and after 5 years exposure under Vietnam weathering condition value.

The geotextile tube supplied was lined with a nonwoven geotextile inner liner filter as fine sand was used for the tube filling (Fig. 9). The nonwoven geotextile filter was stitched to the woven fabric at an interval of 0.3 m centre to centre. This stitching is critical to prevent the delamination of the nonwoven filter geotextile during the pumping operation (Fig. 10). Table 3 lists technical properties of the supplied woven geotextile tube with nonwoven inner liner.

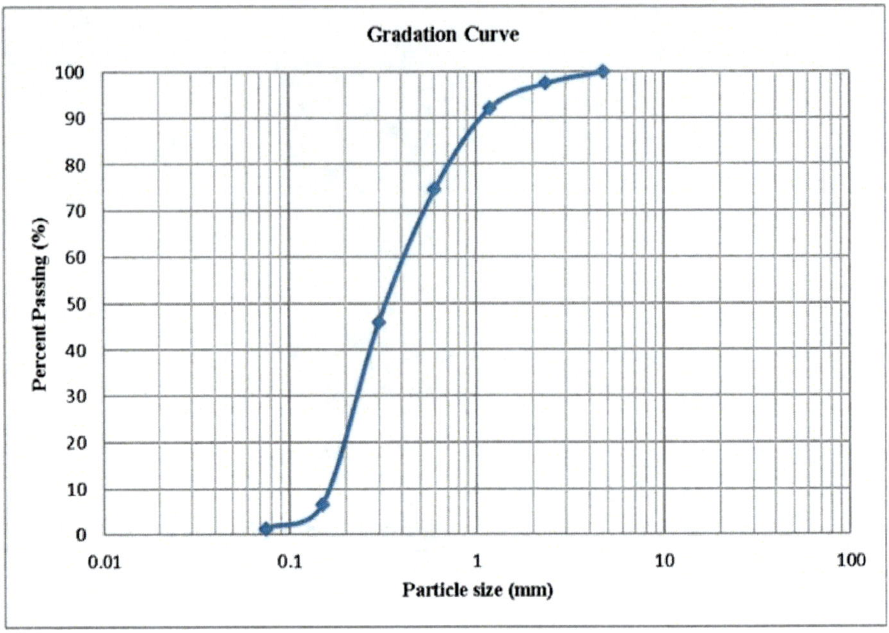

Fig. 9 Gradation curve of the sand used for tube filling

Fig. 10 Geotextile tube with nonwoven filter inner liner

2.4 *High Strength Basal Reinforcement and Sand Filled Mattress*

The external stability analysis was carried out by Slope/W software (by GeoSlope International, Calgary, Canada) using Morgenstern-Price method. Morgenstern-Price Method was used as it used the Moment and Force equilibrium in the computation

Table 3 Technical properties of woven geotextile tube with nonwoven inner liner

Properties	Test standard	Unit	Value
Tensile strength MD	ISO 10319	kN/m	≥200
Tensile strength CD	ISO 10319	kN/m	≥200
Tensile elongation at Ult MD	ISO 10319	%	≤10
Tensile elongation at Ult CD	ISO 10319	%	≤10
CBR puncture strength	ISO 12236	kN	≥22
Pore size O90	ISO 12956	mm	≤0.12
Seam strength CD	ISO 10321	kN/m	≥160

of the Factor of Safety compared to Bishop Method which used only the Force Equilibrium in the Factor of Safety computation [10]. From the external stability required to achieve an F.S. > 1.2, a layer of high strength basal reinforcement geotextile with ultimate tensile strength of 800 kN/m was required at the base of the geotextile tube. The sand filled mattress was designed according to CUR-217 Design Guide and checked against tensile rupture and soil piping [11]. The sand filled mattress was designed to a filled thickness of 180 mm. The fabric used for the sand filled mattress fabrication is made from composite coarse grain that allows soil entrapment, high UV resistance and vandalism resistance.

2.5 Construction Methodology

Scour apron. A layer of bamboo mattress was constructed at a grid formation of 0.7 m × 0.7 m and installed over the soft marine clay. The bamboo mattress allows construction trafficability and also increase the bearing capacity of the soft underlying marine clay. A scour apron of 8.8 m circ was then installed over the bamboo mattress. The scour apron and bamboo mattress are pegged in position using bamboo pegs to ensure that it stays in position during high tide. The scour apron was then pumped to an inflated height of 0.5 m with a width of 3.9 m (Fig. 11).

Geotextile tube. The installation of the geotextile tube had to be properly planned and scheduled to the influence of daily tidal fluctuations. The geotextile tube installation was carried out during the low tide and secured to the timber mattress. A portable steel tank was used for mixing the sand and water to form the sand slurry (Fig. 12). The pumping of sand slurry into the geotextile tube is carried out during low tide by inserting the discharge pipe into "filling ports" at one end of the tube while the other end "filling ports" is left open for water pressure relief. All other intermediate ports are closed. This filling operation is repeated until the whole tube attained the final filling height of 2 m (Fig. 13).

High strength basal reinforcement. The high strength basal reinforcement was supplied in a roll width of 5 m and 100 m length. The roll was then cut to 25 m

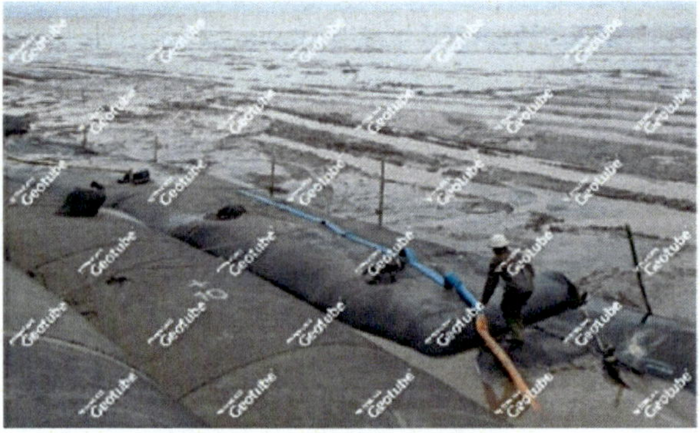

Fig. 11 Scour apron pumped to an inflated height of 0.5 m

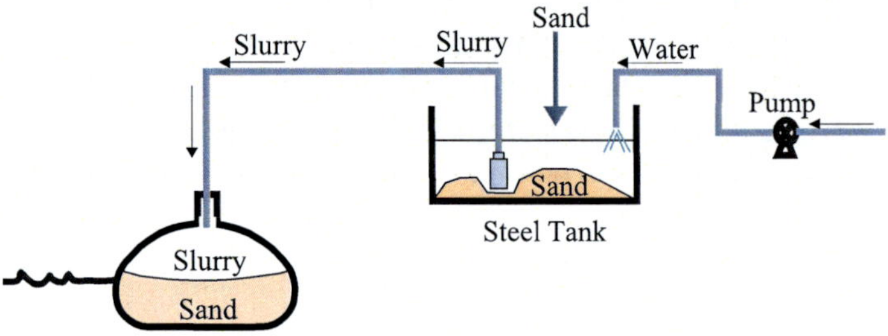

Fig. 12 Schematic of mixing tank and pump set up unit

Fig. 13 Geotextile tube installation in progress

length and seamed to form panels of 16 m × 25 m. The pre seamed reinforcement fabric was then installed over the scour apron during the low tide and pegged into position using bamboo pegs at one meter spacing to prevent movement during sand back filling operation (Fig. 14).

Sand filled mattress. Special composite sand filled mattresses with coarse fibre surfaces were installed on the sloping ground above the crest of the geotextile tubes (Fig. 15). The coarse fibre surface is designed to trap and hold the topsoil in place and allow the growth of vegetation on top. The sand filled mattresses were anchored

Fig. 14 High strength basal reinforcement over bamboo mattress

Fig. 15 Newly installed sand filled mattress at tube crest

into a trench at the crest. The vegetation establishment over the sand filled mattress was very rapid and most of the sand filled mattresses are covered with vegetation after 9 months of exposure (Figs. 16, 17 and 18).

Fig. 16 Sand filled mattress covered with vegetation after 9 months

Fig. 17 Vegetation growth over the sand filled mattress and geotextile tube

Fig. 18 Close up of sand filled mattress covered with vegetation

3 Specially EngineeringComposite Geotextile Tube for Coastal Protection of Meritus Pelangi Beach Resort in Langkawi, Malaysia

3.1 Project Background

The Meritus Pelangi Beach Resort is located at the southwestern coast of Langkawi Island (Fig. 19). Although it is partially sheltered by Pulau Rebak Besar, Pulau Rebak Kecil and Pulau Tepor, it is still susceptible to waves with large fetch lengths from the Andaman Sea. The site is an area of high commercial and tourism value due to the presence of a 5-star resort. The beach is a primary recreational feature for the resort's guests.

Fig. 19 Project location

The conventional sand bags were used for coastal protection but are not effective against erosions. They are displaced, disintegrating when exposed to UV and weathering and are weak against strong wave attacks (Fig. 20). The beach front has narrowed due to relentless waves attack (Fig. 21). To rectify this problem, the Resort evaluated various erosion protection methods. After extensive studies, the solution provided by Tencate with a combination of composite sand colour geotextile tube and geobag was selected due to their overall cost effectiveness and ease of filling and handling (Figs. 22 and 23). Table 4 lists technical properties of the use of composite sand colour geotextile tube.

Fig. 20 Conventional geobags used at site

Fig. 21 A scrap of about 1.5 m high created by wave action on the shoreline

Fig. 22 Composite fabric used for tube fabrication

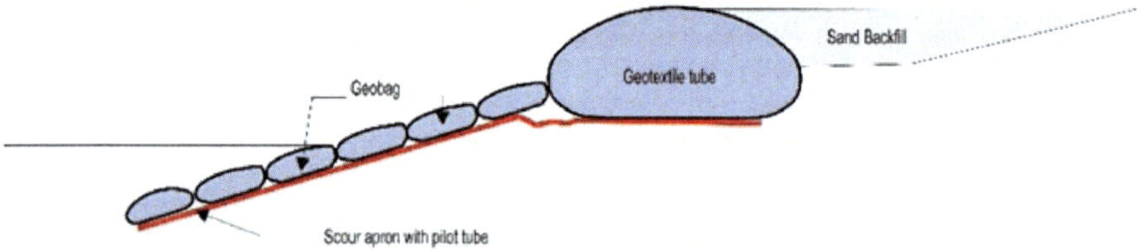

Fig. 23 Cross section drawing of proposed geotextile tube solution

Table 4 Technical properties of composite sand colour geotextile tube

Properties	Test standard	Unit	Value
Tensile strength MD	ISO 10319	kN/m	70
Tensile strength CD	ISO 10319	kN/m	60
Tensile elongation at Ult MD	ISO 10319	%	10
Tensile elongation at Ult CD	ISO 10319	%	10
CBR puncture strength	ISO 12236	kN	9
Pore size O90	ISO 12956	mm	0.2
Seam strength CD	ISO 10321	kN/m	55

3.2 Construction Methodology

Scour Apron. A scour apron was installed with the position of the pilot tube at an 8 m distance from the centre line of the geotextile tube. The scour apron and pilot tube assembly was to protect the foundation of the Geotextile tube unit from scour erosion caused by currents and wave attacks. The pilot tube of the sour apron was then infilled with sand slurry to a designed inflated height of 0.6 m. For most applications, the scour apron assemble should extend from the toe of the geotextile tube unit a minimum of 2.5 times the inflated height of the geotextile tube unit.

Geobag. After the scour apron was installed, Geobags are placed onsite and filled with sand (Fig. 24). The filling ports of each Geobag are then closed using portable stitching machine with polyester thread. The Geobags are lifted and placed using a lifting harness with a spreader beam arrangement (Fig. 25). The Geobags are placed in position on the top of the scour apron by laying in an overlapping shingle style. This is to ensure maximum stability and protection of the shoreline during high water flows.

Geotextile Tube. Once the geobags are fully installed, the contractor positioned the geotextile tube over the scour apron. The pre-stitched loops on the geotextile tube are then attached to wooden poles for position and to ensure the tube does not move during the initial sand pumping operation.

Sand slurry pumps attached to the floating pontoon was used for pumping the sand from the sea into the geotextile tube units (Fig. 26).

Fig. 24 Geobag filling onsite

Fig. 25 Geobag lifting with spreader beam

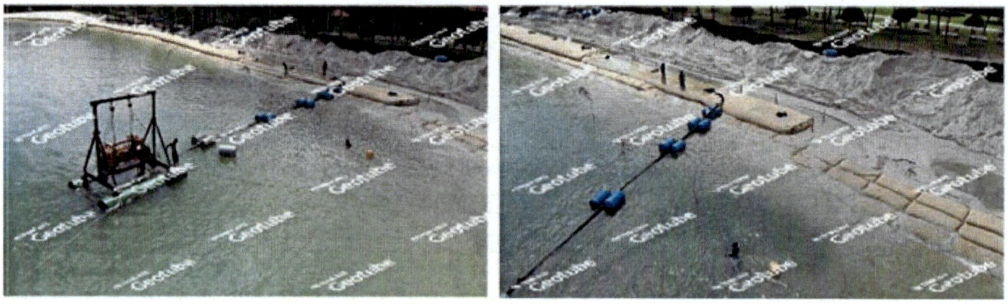

Fig. 26 The sand was hydraulically pumped from the shallow waters direct into the geotextile tube

Sand was then used to cover the geotextile tube that leaves the beach with a natural look. Additional sand was imported to cover the geotextile tube structure and return the beach to its original profile. The beautiful Meritus Beach was successfully restored and protected (Fig. 27).

Fig. 27 The beautiful Meritus beach resort restored and protected

4 Conclusion

This paper presents the various issues related to the geotextile tube design and construction of two case studies of the application of geotextile tubes in coastal protection and land reclamation projects.

The geotextile tubes are properly designed taking into consideration of the internal stability, external stability, survivability and durability of the tube. The fabric used for the geotextile tube should prevent excessive loss of fines but be sufficiently permeable to prevent excessive build-up of pressure during installation. Geotechnically, it must be stable against sliding, overturning, bearing and global slip failure. Also, the tube should endure and perform the engineering functions over the installation process and the whole lifespan of the design whereby the geotextile should be durable in a biological, chemical environment and ultra violet light resistance.

The high strength basal reinforcement, sand filled mattress and geotextile bags are parts of the complete solution to this coastal erosion prevention solution, in addition to the geotextile tubes. All have to be designed taking into account the strength and the environmental considerations.

References

1. Lawson CR (2003) Geotextile containment—international perspectives. In: Proceedings. 17th GRI conference. Geosynthetic Institute, Philadelphia, USA, pp 198–221
2. Nickels H, Heerten G (1996) Building elements made of geosynthetics and sand resist the North Sea surf. In: Proceedings EuroGeo 1, Maastricht, The Netherlands, pp 907–910
3. Artières O, Dunand M, Durand F (2005) Coastal protection with filter tube breakwater "Amelie" beach case. In: Proceedings EuroGeo 3, Munich, Germany, pp 367–370
4. de Bruin P, Loos C (1995) The use of geotubes as an essential part of an 8.8 m high North Sea dyke and embankment, Leybucht, Germany. In: Geosynthetics world, pp 7–10
5. Spelt K (2001) Geotubes as the core of guide dams for Naviduct at Enkhuisen, The Netherlands. Terra et Aqua 83:21–25
6. Fowler J, Ortiz C, Ruiz N, Martinez E (2002) Use of geotubes in Columbia, South America. In: Proceedings 7th international conference on geosynthetics, Nice, France, Balkema, vol 3, pp 1129–1132
7. Yee TW (2002) Construction of underwater dykes using geotextile containment systems. In: Proceedings 7th international conference on geosynthetics, Nice, France, Balkema, vol 3, pp 1161–1164
8. Jackson LA (1987) Evaluation of sand filled geotextile groynes constructed on the Gold Coast, Australia. In: Proceedings 8th Australian conference on coastal and ocean engineering, Launceston, Australia, pp 235–238
9. Palmerton JB (2002) Distinct element modelling of geosynthetic fabric containers. In: Proceedings 7th international conference on geosynthetics, Nice, France, Balkema, vol 3, pp 1021–1024
10. GeoStudio (2012) May 2014 Release version 8.13.1.9253 for slope stability analysis
11. Bezuijen A, Vastenburg EW (2013) Geosystems design rules and applications
12. Yee TW, Zingerink E, Choi JC (2007) Geotextile tube application for Incheon Bridge project, Korea. In: Proceedings of the CEDA dredging days 2007 conference, Rotterdam, Netherlands

13. Yee TW (2015) Dredging works and geotextile tube applications for construction of the longest sea crossing bridge in Vietnam. In: Proceeding of Western Dredging Association and Texas A&M University Center for dredging studies, dredging summit and expo 2015, Texas, USA